Multifractal Theory of Motion

Maricel Agop · Stefan Andrei Irimiciuc

Multifractal Theory of Motion

From Small to Large Scales

Springer

Maricel Agop
Department of Physics
Gheorghe Asachi Technical University
of Iasi
Iaşi, Romania

Stefan Andrei Irimiciuc
Laser Department
National Institute for Laser
Plasma and Radiation Physics
Magurele, Romania

ISBN 978-981-97-6409-9 ISBN 978-981-97-6410-5 (eBook)
https://doi.org/10.1007/978-981-97-6410-5

This Springer imprint is published by the registered company Springer Nature Singapore Pte Ltd.
The registered company address is: 152 Beach Road, #21-01/04 Gateway East, Singapore 189721, Singapore

Preface

An important aspect is that the constraints of determinism do not imply regular behavior or do not rely on predictability in describing the dynamics of natural systems. Currently, exploring the evolution of natural systems has had a great impact on a wide range of scientific domains. In the linear analysis on which scientific development was based before the seventh decade of the twentieth century, unlimited predictability was automatically associated with natural systems. The development of nonlinear analysis and the discovery of the scientific foundations of chaos have radically changed the view of scientific analysis by demonstrating that the reductionist analysis method on which entire scientific domains were developed has limited implementation and that the unlimited predictability of systems is not a characteristic of real systems but an approximation resulting from linear treatment.

The dynamics of populations, cardiac fibrillation, communication systems, fluid flow, and atmospheric systems are just a few topics in which nonlinearity and chaos are regular manifestations. This leads to the idea that it is necessary to find universal laws that govern these dynamics. The variety of scientific topics has also attracted increasing interest in recent decades from readers in both the scientific community and the nonscientific community.

This book is dedicated to nonlinear and chaotic dynamics, a domain that breathes trans- and multidisciplinary and has been the focus of the scientific community in recent decades. The implementation of operational procedures (group invariants, differentiable geometries, variational principles, harmonic mapping, etc.) will provide a background for the development of a new multifractal theory of motion for the study of natural system dynamics. We would like to mention that in the author's view, such a theory would act as an extension of the Scale Relativity Theory proposed by Nottale for scaling the fractal dimension (from $D_F = 2$, as it was originally proposed). The development of a universal approach to describing the overall dynamics laws of natural systems will manifest itself through the complex potential method at various resolution scales (the complex potential of the gravitational field in a Newtonian sense, the complex potential of the gravitational field in an Einsteinian sense—Ernst potential—the velocity complex field in the Nottale sense), or through the harmonic mapping method implemented at various resolution scales. Basically, natural system dynamics will be projected at various resolution scales, from atomic to cosmological.

The book aims to present this topic in a manner that would be useful and beneficial to students and young researchers from various fields such as physics, mathematics, biology, or even medicine.

Iaşi, Romania Maricel Agop
Magurele, Romania Stefan Andrei Irimiciuc

Contents

Chapter 1
From Small to Large Scales Through the Dynamics of Kepler-Type Problems

1.1 Dynamics of Kepler-Type Problems

The classical Kepler motion is the model depicting the revolution of planets around the Sun, or the revolution of the electrons around the nucleus, within the framework of classical dynamics. The image is usually rendered with reference to material points taken as pure positions, and can be dynamically explained with the help of Newtonian equations of motion. In vector notation, these are

$$\ddot{\boldsymbol{r}} + \frac{K}{r^2}\frac{\boldsymbol{r}}{r} = \boldsymbol{0}. \tag{1.1}$$

Here K is a constant, **r** denotes the position vector, with respect to the center of force, of the material point whose motion is calculated, and a dot over a symbol means derivative with respect to time, as usual. The constant K does not depend on quantities related to the point considered in motion, but only in cases where electric forces are involved. We can simplify the algebra leading to the solution of Eq. (1.1) by restricting the geometry to the plane of motion, as a benefit of the vector expression of the centrality of force. If the generic coordinates of the point in motion are ξ and η say, the Eq. (1.1) is then equivalent to the system [1]

$$\ddot{\xi} + K\frac{\cos\phi}{r^2} = 0, \ \ddot{\eta} + K\frac{\sin\phi}{r^2} = 0 \tag{1.2}$$

with r and ϕ polar coordinates of the moving point, in the plane of motion, with respect to the attraction center. The magnitude of the rate of area swept by the position vector is then given by

$$\dot{a} \equiv \xi\dot{\eta} - \eta\dot{\xi} = r^2\dot{\phi}. \tag{1.3}$$

M. Agop and S. A. Irimiciuc, *Multifractal Theory of Motion*,
https://doi.org/10.1007/978-981-97-6410-5_1

This constant of motion allows us an elegant integration of the system (1.2) with the analytical form of the trajectory as a direct outcome. First, we define the complex variable

$$z \equiv \xi + i\eta = re^{i\phi} \tag{1.4}$$

so that (1.2) can be written in the form

$$\ddot{z} + \frac{K}{r^2}e^{i\phi} = 0. \tag{1.5}$$

Now, use (1.3) to eliminate r^2, such that

$$\ddot{z} + \frac{K}{\dot{a}}e^{i\phi}\dot{\phi} = 0 \;\therefore\; \dot{z} = i\left(\frac{K}{\dot{a}}e^{i\phi} + w\right), \tag{1.6}$$

where $w \equiv w_1 + iw_2$ is a complex constant of integration, to be obviously determined by the initial conditions of the problem. The analytical equation of motion can be extracted then directly, by calculating the area constant (1.3) with the help of the first result of integration given in Eq. (1.6). In polar coordinates, the result is given in the form:

$$\frac{\dot{a}}{r} = \frac{K}{\dot{a}} + w_1 \cos\phi + w_2 \sin\phi. \tag{1.7}$$

The shape of this trajectory is best pictured by going back to the Cartesian coordinates ξ and η, , whereby we have, instead of (1.7) the second-degree curve a conic:

$$\left(\frac{K^2}{\dot{a}^2} - w_1^2\right)\xi^2 - 2w_1w_2\xi\eta + \left(\frac{K^2}{\dot{a}^2} - w_2^2\right)\eta^2 + 2\dot{a}(w_1\xi + w_2\eta) = \dot{a}^2. \tag{1.8}$$

The center of this conic is not the center of the force, i.e., the origin of the plane of motion in our case, but has the coordinates

$$\xi_0 = -\frac{\dot{a}w_1}{\Delta},\ \eta_0 = -\frac{\dot{a}w_2}{\Delta};\ \Delta \equiv \left(\frac{K}{\dot{a}}\right)^2 - w_1^2 - w_2^2. \tag{1.9}$$

In cases where $\Delta = 0$, the center of this trajectory is at infinity: the trajectory is a parabola.

Assuming the center of the trajectory at a finite distance with respect to the center of force, and referring the trajectory to this center by a translation: $x = \xi - \xi_0$, $y = \eta - \eta_0$, its equation becomes

$$\left(\frac{K^2}{\dot{a}^2} - w_1^2\right)x^2 - 2w_1w_2xy + \left(\frac{K^2}{\dot{a}^2} - w_2^2\right)y^2 = \frac{K^2}{\Delta}. \tag{1.10}$$

The quadratic form from the left-hand side of this equation is completely characterized by the 2 × 2 special matrix

$$\mathrm{a} \equiv \begin{pmatrix} \frac{K^2}{\dot{a}^2} - w_1^2 & -w_1 w_2 \\ -w_1 w_2 & \frac{K^2}{\dot{a}^2} - w_2^2 \end{pmatrix}. \tag{1.11}$$

This matrix is typical for guiding our reason: its entries contain both physical and initial conditions. The eigenvalues of this matrix are Δ and $K^2/\dot{a}^2$, with the corresponding eigenvectors

$$|e_1\rangle = \begin{pmatrix} \cos\omega \\ \sin\omega \end{pmatrix}, |e_2\rangle = \begin{pmatrix} -\sin\omega \\ \cos\omega \end{pmatrix}; w_1 = |w|\cos\omega, w_2 = |w|\sin\omega. \tag{1.12}$$

Thus, the orientation of the trajectory in its plane is defined by the initial conditions of the motion which, moreover, also properly define its relative asymmetry in the two directions of the axes. Specifically, the asymmetry is measured by the ratio of the two principal dimensions of the ellipse, which boils down to the fact that the magnitude $|w|$ is proportional to the eccentricity " e" of trajectory. Indeed, the semiaxes, customarily denoted by "a" and "b", and the eccentricity "e" calculated with their values, are given by

$$a^2 = \frac{K^2}{\Delta^2}, b^2 = \frac{\dot{a}^2}{\Delta} \therefore e^2 \equiv \frac{a^2 - b^2}{a^2} = \left(\frac{\dot{a}}{K}\mathbf{w}\right), \tag{1.13}$$

where $\mathbf{w}$ is the vector equivalent of the complex number "w". This notation makes the vector nature of the eccentricity obvious. Therefore, the initial conditions can actually be expressed only in terms of some "contemporary" magnitudes, to which we have only to add the partially controllable arbitrariness of an angle:

$$w_1 = \frac{K}{\dot{a}} e \cdot \cos\omega, w_2 = \frac{K}{\dot{a}} e \cdot \sin\omega. \tag{1.14}$$

This allows us, to a great extent, to forget, at least partially, about the past: as long as it is represented by the initial conditions, it can be found in contemporary measurable quantities. These are, obviously, the components of the vector $\mathbf{w}$ from Eq. (1.14). This is essentially the observation that prompted the Newtonian explanation for the real planetary motions in terms of contemporary quantities. Indeed, a force is always contemporary, and the initial conditions of the motion whatever they might have been in a remote past which is never within our reach are then to be read, at least partially, in some contemporary parameters of motion, specifically the area constant and the eccentricity.

1.2 The Geometry Associated to Dynamics of Kepler-Type Problems and Complex Potential

As one can see directly from Eqs. (1.13) the semi axe "b" can be imaginary for $\Delta < 0$, in which case we have to do with hyperbolic trajectories. It is only in cases where $\Delta > 0$, that we have to do with elliptic trajectories, properly depicting the planetary motion. Along this line of reasoning, the parabolic trajectories are all characterized by points on the circle $\Delta = 0$, i.e.

$$\mu^2 + \nu^2 = 1,\ \mu \equiv e \cdot \cos\omega,\ \nu \equiv e \cdot \sin\omega. \tag{1.15}$$

Then the whole interior of this circle corresponds to all possible finite Kepler motions that a material point can have around a center acting with a force inversely proportional to the square of distance. This would mean that the motion of a planet or an electron indicates infinitely many possible initial conditions, from which it would have had to "choose" so to speak. The fact is though, that the actual motion of a planet is perceived as if it would have unique initial conditions. Any departure from this perception has always induced arguments about some actual perturbations acting on the planet. To a certain extent, this is true: the discovery of Neptune is an example. However, as history shows, it has not that touch of universality needed for the continuity of our knowledge and especially for cosmologic and cosmogonic purposes. Thus, we have to turn to the origin of the problem, and lead our reasoning along the following lines: Kepler motion should have reality only as a "snapshot", this is an undeniable fact, for it could not have been discovered otherwise. However, the planetary motion, even if we confine its description to a Kepler motion, is a succession of such snapshots, which have to be put together in order to make the whole thing. First of all, we have to find the time scale of such a snapshot, and that is hard. But we have another possibility, opened by the remarks just made before. Namely, there is an a priori metric geometry for the defining parameters of the snapshot, which are in fact the initial conditions of the dynamical problem describing it. This geometry defines in turn a kinematics, and that kinematics offers us a natural way to continuously connect the snapshots in a succession representing a real trajectory.

Indeed, even superficially, it can be seen at once that the above-mentioned freedom of the parameters defining the types of Kepler orbits, allows us to construct a Cayley-Klein (or, as we called it before, absolute) geometry [2, 3] characterizing the variation of those orbits. We know that an absolute geometry is related to some conservation laws, at least as long as some realizations of SL(2, R) group structure are involved. And indeed, the absolute metric for the interior of the circle (1.15)

$$(ds)^2 = \frac{\left(1-\nu^2\right)(d\mu)^2 + 2\mu\nu(d\mu)(d\nu) + \left(1-\mu^2\right)(d\nu)^2}{\left(1-\mu^2-\nu^2\right)} \tag{1.16}$$

can be brought to the form of Poincaré metric

$$(ds)^2 = -4\frac{dh \cdot dh^*}{(h - h^*)^2} = \frac{(du)^2 + (dv)^2}{v^2} \tag{1.17}$$

by the following transformation of coordinates:

$$\mu = \frac{hh^* - 1}{hh^* + 1},\ v = \frac{h + h^*}{hh^* + 1} \leftrightarrow \begin{matrix} h \equiv u + iv = \frac{v + i\sqrt{1 - \mu^2 - v^2}}{1 - \mu}. \\ h^* = u - iv \end{matrix} \tag{1.18}$$

The conservation laws for the metric (1.17) are represented by the following differential 1-forms:

$$\omega_1 = \frac{du}{v^2},\ \omega_2 = 2\frac{udu + vdv}{v^2},\ \omega_3 = \frac{(u^2 - v^2)du + 2uvdv}{v^2}. \tag{1.19}$$

A natural description of the Kepler motion is the one in variables (e, ω), i.e., the eccentricity and the orientation of the orbit in its plane. In terms of these parameters, the metric (1.16) becomes

$$(ds)^2 = \left(\frac{de}{1 - e^2}\right)^2 + \frac{e^2}{1 - e^2}(d\omega)^2. \tag{1.20}$$

We can rewrite this metric in a well-known form, by recalling that for elliptic trajectories " e" is confined to the interval between -1 and $+1$, so that the change of parameter

$$e = \tanh \psi \tag{1.21}$$

is legitimate. With this the metric, (1.20) becomes

$$(ds)^2 = (d\psi)^2 + \sinh^2\psi(d\omega)^2. \tag{1.22}$$

The complex parameter "h" from Eq. (1.18) has now a direct connection with the theory of classical Newtonian potentials via a harmonic map. In order to show this relationship, we write here "h" from Eq. (1.18) in terms of (e, ω). We have

$$h = i\frac{\cosh \chi + \sinh \chi \cdot e^{-i\omega}}{\cosh \chi - \sinh \chi \cdot e^{-i\omega}},\ \chi \equiv \frac{\psi}{2} \tag{1.23}$$

or explicitly

$$h = \left(\frac{2r \sin[2\Omega(t - t_r)]}{1 + r^2 + 2r\cos[2\Omega(t - t_r)]} + i\frac{1 - r^2}{1 + r^2 + 2r\cos[2\Omega(t - t_r)]}\right),$$

where

$$r = \coth \chi, \ \omega = 2\Omega(t - t_r).$$

As it happens, this equation represents a harmonic map from the usual space into the Lobachevsky plane having the metric (1.17), provided χ (and therefore ψ) is a solution of the Laplace equation in free space.

In these figures, Real (h) $\equiv F(\omega, t)$ (the amplitude at various scale resolutions given by the maximum value of Ω) is represented as functions of t and Ω for r = 0,5.

As can be observed in Figs. 1.1a–d and 1.2a–d, the natural transition of gravitational system dynamics is to evolve from a normal period doubling state towards modulated dynamics. The gravitational system dynamics never reach a chaotic state but they permanently evolve towards that state.

Indeed, the problem of harmonic correspondences between space and the hyperbolic plane is described by the stationary values of energy functional corresponding to the metric (1.17). This is defined as the volume integral of an integrand obtained from that metric by transforming the differentials into space gradients [4, 5]. The stationary values of the energy functional, therefore, correspond to solutions of the Euler–Lagrange equations for a Lagrangian-like

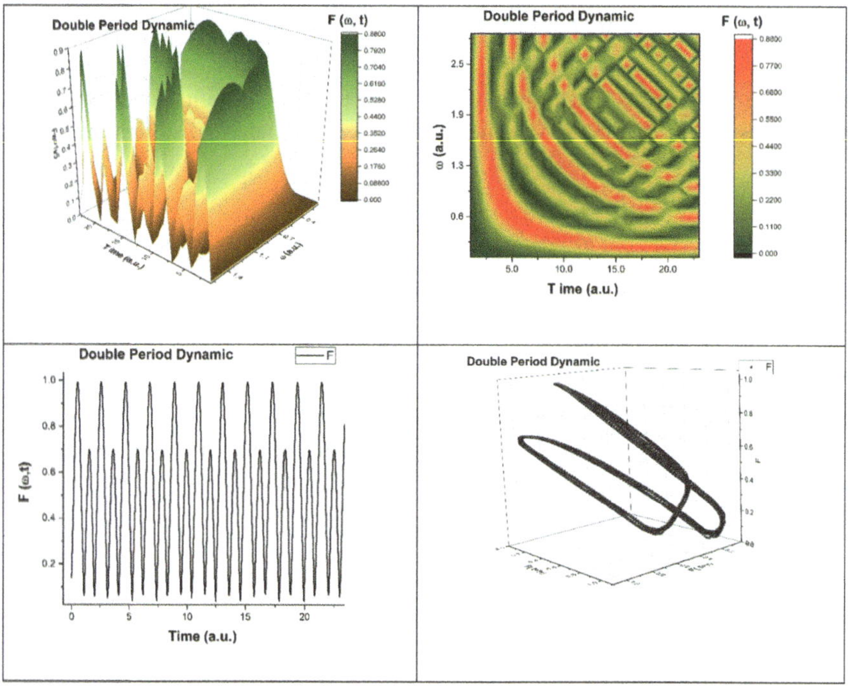

Fig. 1.1 a–d: The "double period dynamic modes" of the gravitational system dynamics are presented: **a** 3D diagram, **b** contour diagram, **c** time series, and **d** reconstituted attractor for scale resolutions given by Ω_{max}

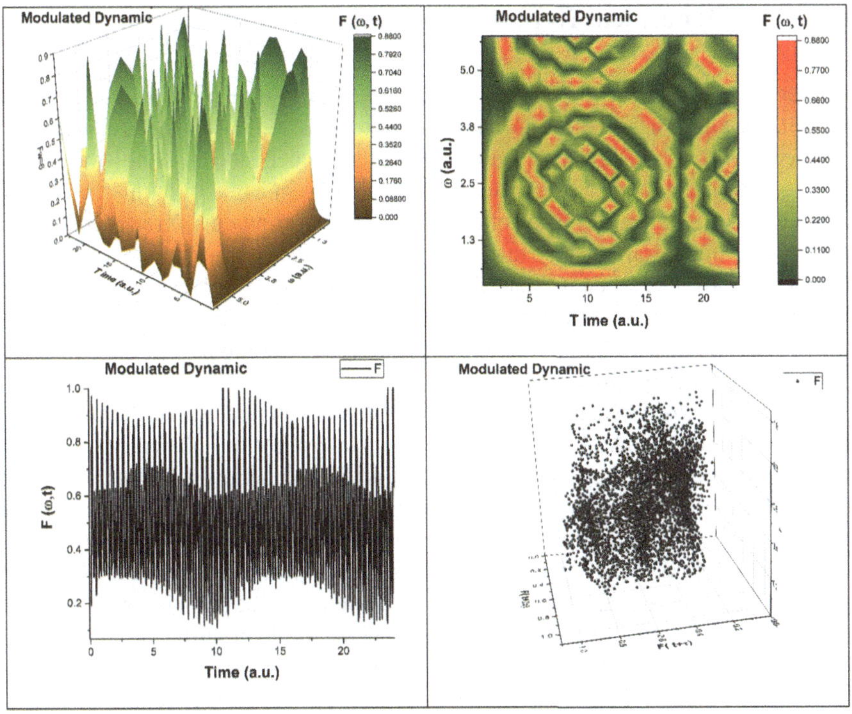

Fig. 1.2 **a**–**d**: The "modulated dynamic modes" of the gravitational system dynamics are presented: **a** 3D diagram, **b** contour diagram, **c** time series, and **d** reconstituted attractor for scale resolutions given by Ω_{max}

$$\Lambda \equiv -4\frac{\nabla h \cdot \nabla h^*}{(h - h^*)^2}. \tag{1.24}$$

These are

$$\left(h - h^*\right) \cdot \nabla^2 h - 2(\nabla h)^2 = 0 \tag{1.25}$$

and its complex conjugate. Then it is easy to see, by a direct calculation, that h from (1.23) verifies this Eq. (1.25) when χ is a solution of the Laplace equation, and ω is arbitrary, in the sense that it does not depend on the position in space. Nevertheless, it might depend on the local time of the Newtonian dynamics. We note that the quantities hand $\overline{h}$ can be put in correspondence with Ernst's complex potential from General Relativity [6, 7].

1.3 The Dynamical Kepler-Type Problems and Skyrmions

The development at the confluence of geometry and dynamics, physics has found that the planetary motion in the case of the solar system, as well as the electronic motion in the case of the classical atomic model, can be fairly depicted as a Kepler problem. This condition, let us recall once again, is rigorously satisfied only in cases where the bodies are point-like. The vector representing eccentricity in a Krettooepler orbit is related to the initial velocity that a planet, or an electron in the case of atom, is assumed to have had when it "started" orbiting around the Sun, respectively around the atomic nucleus. One can rightfully say that by means of the present measurements of the eccentricities, we have at least a "partial" access to the past of the planets or electrons, as the case may occur.

One of the most important conclusions of this perception is that the indecision in the initial conditions of the Kepler motion describing the planetary or atomic dynamics, has a precise geometrical form: it is the geometry of the interior of the unit circle, therefore the hyperbolic geometry of Lobachevsky. As in the case of planets these eccentricities are small, in real space terms the hyperbolic geometry applies only to a small space region, practically concentrated to the volume of the Sun itself. The same can be said, by analogy, of the nucleus—if the atom is discussed. It is, therefore, conceivable that this hyperbolic geometry would describe the content, and the manner of producing, of both solar and nuclear energy, by a kind of deformation process of a continuous structure of their matter. This is actually what Lord Kelvin tried to explain by the middle of the nineteenth century, regarding the source of solar energy.

Here the most important conclusion though, is that this method can be thought of as really ascribing "a space expanse" of the nuclear matter, in the form of harmonic surfaces in space, to the regions of space spreading over ranges whose measure is the eccentricity of the Kepler motion. In other words, the harmonic maps from the physical space to the Lobachevsky plane are intimately related either to the physics of the Sun, in the case of the planetary system, or to the physics of the nucleus in the case of the classical atomic model. There is no obvious sign, in physics today, of some such theoretical description in the first case, i.e., the case of the Sun. That is, if we don't consider the contemporary preoccupations with the seismicity of the Sun and planets, which, from some points of view, may qualify as an application of harmonic mappings. However, such a conclusion seems to be fair in the case of the atomic model, where the nuclear matter can be described by harmonic maps, and more often at that lately, by the model of Skyrme.

Before getting into that subject, let us make nevertheless a connection with the conclusions from the first chapter, regarding the space expanse of the atomic nucleus. In the description above, the central expanse of the atomic space plainly qualifies as the space into which the electron plunges, or from which it is ejected, in some classical "Wilson processes" that involve or not a production of light. In other words, if the condition of the material point of the nucleus is the one making us reluctant to accept the reality of such processes, then it is time to forget about it. Indeed, the contemporary space expanse of the atomic nucleus, like the contemporary space expanse of the Sun itself for that matter, should be the reflection into the present of a distant past, as it appears in the description of the planetary system by Kepler motions.

This fact puts an interesting spin—properly, as well as figuratively, as we will see here–on the ejection of an electron or its absorption by the nucleus. When plunging into the nucleus, an electron appears as if it goes into the past, while when it is ejected from the nucleus it suddenly pops up into the present. The past and the present of an electron meet, therefore, inside the nuclear matter. For once, part of this process—popping up into the present—has been met in a popular model involving the macroscopic counterpart of the atom—the planetary system. Indeed, the most familiar cosmogonic model of the planetary system is that in which the planets are born from a primary nebula by a process of ejection due to the pure mechanical effect of the rotation. By carrying the analogy in reverse—i.e., from the atomic model to the planetary one—one might say that the cosmogonic model misses an essential part, corresponding to the ingestion of the electron by the nucleus. Indeed, it is momentarily hard to say, within astrophysical experience today, what would be a cosmogonic process that corresponds to this quite natural one from the atomic realm.

In hindsight, this transition present-past and vice versa means to offer a physical content to time. The fact was brought to light for the first time by Richard Feynman in the construction of his famous graphs [8, 9]. These represent the transition from the usual continuous description in space and time by differential equations, to the apparent accidentality of events in the atomic realm. They make obvious the necessity of discussing the time by its physical content. The positron, for instance, is an electron going back in time: while for the electron the time goes normally, toward the future, for the positron the time goes toward the past.

One other pertinent observation here is related to the metric (1.22). It is well known in theoretical physics: the Fock metric of the velocity space in special relativity [10]. Indeed, this metric can be produced as an absolute metric in the three-dimensional space of velocities, in view of the fact that all possible velocities of the matter formations in the universe are smaller than the speed of light. The resulting metric is given by Eq. (1.20) where e is this time the ratio of the velocity in matter and the speed of light, and ω is the metric on the unit sphere. One can say that the only thing well specified here is the speed of light. All the other velocities, regarding material bodies are contingencies. This is why the Fock metric is well suited to producing a priori minimum information measures in velocity space [11]. The metrics (1.20) or (1.22) provide just such an a priori minimal measure, based on the same principles as the relativistic ones. As a matter of fact, their very source, the absolute metric (1.16) involves the special metric tensor, characteristic of a constant curvature space. This conveys properties of universality to a special geometrization of the harmonic maps representing the nucleus.

1.4 The Geometry Associated to the Skyrmions

In the following, we show that the Skyrme theory is naturally suited for the nuclear matter, inasmuch as it is simply a variation on the harmonic maps. Conceived initially for mapping Euclidean structures, it had to accommodate the soliton-like constitution of matter by nonlinear terms in the energy functional. Then, in order to describe the baryons, which should be at home within the nucleus the energy functional had to include a term pertaining to a higher cosmology class. From this point of view, Skyrme's functional had to be algebraically non-homogeneous. The Manton's geometrization gives us an opportunity to "homogenize" it so to speak.

Variation on the Theme of Harmonic Maps

We don't need to insist too much in recalling that, in spite of many significant shortcomings, the main scientific image of the atom was always, in the last hundreds of years as well as today, the intuitive planetary one, amended perhaps with the idea that such a physical system is not plane but a spatial one—maybe spherical. This sphericity of the model, if real, should be due either to the non-centrality of forces dynamically responsible for determining the Kepler motion or, as we shall endeavor to explain here in the limits of the Newtonian theory of forces, to the space expanse of matter, at the electronic and nuclear level. If this is the case, then as we have shown in the previous chapter, the region of the nucleus can be characterized by a three-dimensional hyperbolic space related to the regular Euclidean space of our experience by harmonic mappings. With this statement, we enter the realm of the recent date of the hyperbolic skyrmions [12, 13].

Indeed, one of the dominant contemporary concepts in the theory of the structure of the nuclear matter, is that of skyrmion, which is but a variation about harmonic applications. The skyrmion is a topological soliton that can represent, in appropriate conditions, the nucleons, as forms of the nuclear matter. The history of this subject begins with the physicist Tony Skyrme, and the reader in need of following it closely can start with the very recollections of Skyrme on the beginnings [14, 15], where the reasons, and the original works, are to be found. The main mathematical point of Skyrme's idea though, is a certain—"almost harmonic" we should say—map from the usual space to the sphere of the relativistic space, built after the traditional manner of the variational problem leading to the Laplace and Schrödinger equations. Let us briefly describe this manner, following, by and large, the guidance of the recent work of Slobodeanu [16], however with the focus on maps from the usual space to the hyperbolic space, or hyperbolic plane. The reason is quite clear: based on our previous considerations, we feel that such maps legitimately describe the nuclear matter filling the central space of a classical atom. The "filling" itself, we are talking about here, is actually represented, as we have seen before, as a harmonic map, or a variation thereupon.

As known, such a harmonic map is obtained by finding the functions that realize the extremum of an energy functional of the form

$$E_2(\boldsymbol{\phi}) = \iiint \langle d\boldsymbol{\phi}|d\boldsymbol{\phi}\rangle d^3\boldsymbol{x}, \tag{1.26}$$

where $\mathbf{d^3X}$ is the volume element of space. Here the function realizing the correspondence is, generally speaking, of the nature of a matrix, and this is why we denoted it by a bold $\boldsymbol{\Phi}$. . Its differential actually represents a space gradient of the correspondence in the current metric of the ambient physical space—the Euclidean metric space. On the other hand, the internal product from the functional (1.26) is the one induced by the metric of the hyperbolic plane or space. In the case of Kepler motion, for instance, we may write

$$\begin{aligned} \phi \equiv \begin{pmatrix} \boldsymbol{h}+\boldsymbol{h}^* & -2\boldsymbol{h}\boldsymbol{h}^* \\ 2 & -\boldsymbol{h}-\boldsymbol{h}^* \end{pmatrix}; \langle d\phi|d\phi\rangle \\ \equiv -4\frac{\nabla\boldsymbol{h}\cdot\nabla\boldsymbol{h}^*}{\left(\boldsymbol{h}-\boldsymbol{h}^*\right)^2}, \end{aligned} \tag{1.27}$$

where "h" is the complex number defined earlier, comprising the eccentricity and the orientation of the orbit, and h^* is its complex conjugate. In other words, the Kepler motion is represented by the matrix of its quadratic form, written in terms of its asymptotic directions. In the general case, the integrand in Eq. (1.26) has the form [5]

$$\langle d\boldsymbol{\phi}|d\boldsymbol{\phi}\rangle \equiv g_{\alpha\beta}\frac{\partial\phi^\alpha}{\partial\boldsymbol{x}^i}\frac{\partial\phi^\beta}{\partial\boldsymbol{\phi}^j}h^{ij}, \tag{1.28}$$

where the summation convention applies to all indices. Here $\mathbf{g}$ is the metric in the space of mappings $\boldsymbol{\Phi}$, whose components are indexed by Greek indices, and $\mathbf{h}$ is the metric in the ambient space, for which we use the Latin indices. In the cases where $\mathbf{h}$ is Euclidean and $\mathbf{g}$ is the Poincaré metric, we have the expression of the integrand as written in Eq. (1.27).

The original theory of Skyrme doesn't take this "already built-in" aspect of the classical model of the atom into consideration. Rather, the nucleus is assumed to have a particulate structure in the regular Euclidean 4D space, and in this case, the harmonic principle leads to a linear equation. Indeed, in this case, we have to do with the Euclidean 3-sphere and the mapping $\boldsymbol{\Phi}$ is given by the normal vector of the unit sphere, $\boldsymbol{n}$ say. Therefore, Eq. (1.28) takes the form

$$\langle d\boldsymbol{\phi}|d\boldsymbol{\phi}\rangle \equiv \delta_{\alpha\beta}\frac{\partial n^\alpha}{\partial x^i}\frac{\partial n^\beta}{\partial x^j}\delta^{ij} = \sum\nolimits_\alpha (\nabla n^\alpha\cdot\nabla n^\alpha). \tag{1.29}$$

Thus, the differential equation corresponding to the functional (1.26) is linear: the vector $\boldsymbol{n}$ has as components solutions of the Laplace equation. In view of this linearity of the resulting equation, if not somehow amended, the harmonic principle cannot account for the particulate structure, which here would occur in the classical form of topological solitons. So, it appears that the Skyrme's original functional, which was formulated in terms of deformations of a regular Euclidean space [17], is not quite as simple as the above one, but involves higher degree terms, belonging to a different cosmology class, and given by the equation

$$E_4(\boldsymbol{\phi}) = \iiint \left\langle d\boldsymbol{\phi}^\wedge d\boldsymbol{\phi} \middle| d\boldsymbol{\phi}^\wedge d\boldsymbol{\phi} \right\rangle d^3\boldsymbol{x}, \tag{1.30}$$

where "∧" is the exterior product of the differentials. Incidentally, the indices 2 and 4 occurring in the energetic functionals of Skyrme are justified by the orders of the tensors entering the integrands of those functionals. In this case, we must first construct the 2-forms entering the interior product. They are defined according to the matrix multiplication of d**Φ**, so that in the previous situation, we will have

$$d\boldsymbol{\phi}^\wedge d\boldsymbol{\phi} \equiv \nabla n^\alpha \times \nabla n^\beta. \tag{1.31}$$

This time, however, one has to do with a skew-symmetric tensor in the upper indices, and the internal product from Eq. (1.30) is largely a matter of choice. The usual choice is the trace of the square of the tensor:

$$\left\langle d\boldsymbol{\phi}^\wedge d\boldsymbol{\phi} \middle| d\boldsymbol{\phi}^\wedge d\boldsymbol{\phi} \right\rangle = \sum_{\alpha\beta} \left(\nabla n^\alpha \times \nabla n^\beta\right)^2. \tag{1.32}$$

Thus, the actual Skyrme functional corresponding to a situation describing the nuclear matter, was originally written in algebraically non-homogeneous form, something like

$$E(\boldsymbol{\phi}) = \frac{1}{2} \iiint \left(a\langle d\boldsymbol{\phi} | d\boldsymbol{\phi}\rangle + b\left\langle d\boldsymbol{\phi}^\wedge d\boldsymbol{\phi} \middle| d\boldsymbol{\phi}^\wedge d\boldsymbol{\phi}\right\rangle\right) d^3\boldsymbol{x} \tag{1.33}$$

and is referring to a mapping **Φ** from the real space to itself. This one is now determined as a solution of the nonlinear differential equation resulting from the variational principle which refers to the functional (8). This equation is

$$a\nabla^2 n^\alpha + n_\beta\left(n^\alpha \times n^\beta\right) = 0. \tag{1.34}$$

In a particular case, Skyrme presents a "static solution" in the form of the "hedgehog ansatz", representing a nucleon

$$\phi(\boldsymbol{r}) = \exp\{(\boldsymbol{\tau} \cdot \hat{\boldsymbol{r}})\boldsymbol{F}(r)\}, \tag{1.35}$$

where $\boldsymbol{\tau}$ is the triple of Pauli matrices representing the nuclear isospin. The Eq. (1.33) then turns actually into a differential equation for the function F(r), which can be solved in different assumptions. Almost any variation on the Skyrme model, including the original one of Skyrme, who introduced the baryonic number as a topological invariant, takes this basic situation for granted. Exceptions are a few approaches which, so much the more, reflect the indecision of the model outside the realm of the mappings from ordinary Euclidean space to itself.

There are, however, works that touch our own subject matter here, and quite explicitly at that. One of these seems, for the moment, worth mentioning again, for it guides our guesses. In their work from 2004, Atiyah and Sutcliffe [13], seem to point out the fundamental significance of hyperbolic skyrmions. A conclusion of this work is indeed especially noticeable, namely that the theory of Euclidean skyrmions with massive pions leads to detailed calculation results which are almost identical with those referring to massless hyperbolic skyrmions. It is as if the mass—the source of inertia in classical mechanics and general relativity—is somehow related to the (constant) curvature this way. This is, again, quite a fundamental idea of general relativity, however carried over not in space time but simply in space.

It is thus symptomatic that a structure of space per se, which is almost of no concern in the like problem of general relativity, for instance, kicks back into a theory that seems so far away from it. In this respect, it is perhaps worth recalling that the vacuum or electrovacuum Einstein equations are equivalent to the harmonic principle represented by the metric from Eq. (1.27). The resulting differential equation is actually "already" nonlinear, leading directly to solitons [18]. It is this aspect of the theory of hyperbolic skyrmions which, when combined with the results of Atiyah and Sutcliffe, seems to lead to a promising theoretical understanding of the nuclear matter. From this perspective, it seems just a mishap that the theory of Euclidean skyrmions appears to be completely disjoint from that of the hyperbolic skyrmions. However, it is indeed just an appearance, inasmuch as, going along the line of Euclidean skyrmions, one of the most notable results has emerged, that gives us the possibility to put the hyperbolic skyrmions on a sound physical basis, on a par we might say, and with the very help of the Euclidean skyrmions.

Theory of Manton: A Matter Bridge Between Geometries

The exquisite work of Nicholas Manton [19] is referring indeed to Euclidean skyrmions. He, however, presents a "natural" geometrization of them whereby the skyrmions are related to deformations of matter. In fact, nothing seems more natural: the nuclear matter is, after all, just matter, and as the Nature around us suggests it, the most general physical process that matter, in any of its instances, can undergo is a deformation. Besides, there is a question of convenience here, for the most successful description of the deformation of matter is the one that takes the matter as a continuum, and this happens to fit our inherent lack of knowledge of the structure of nuclear matter.

Manton's idea is distinctively one of a classical inspiration, taken indeed from the deformation theory, where the experimental deformations are described by the so-called tensor of stretches [20, 21]. The most general energetic functional of Manton can be written in the form given in Eq. (1.33), where $\mathbf{\Phi}$ is now a vector dictated by the deformation of matter in the regular space:

$$\phi(x, y, z) = \begin{pmatrix} \lambda_1 & 0 & 0 \\ 0 & \lambda_2 & 0 \\ 0 & 0 & \lambda_2 \end{pmatrix} \begin{pmatrix} x \\ y \\ z \end{pmatrix} \tag{1.36}$$

with $\lambda_{1,2,3}$ the elongations along three reciprocally orthogonal directions (the stretches). Here the inner product from (1.33) is simply the dot product of regular vectors. As from Eq. (1.36) we have

$$d\phi = \begin{pmatrix} \lambda_1 & 0 & 0 \\ 0 & \lambda_2 & 0 \\ 0 & 0 & \lambda_2 \end{pmatrix} \begin{pmatrix} dx \\ dy \\ dz \end{pmatrix} \tag{1.37}$$

it is plain that the energy functional from Eq. (1.26) is

$$E_2(\boldsymbol{\phi}) = \iiint \left(\lambda_1^2 + \lambda_1^2 + \lambda_3^2\right) d^3\boldsymbol{x}. \tag{1.38}$$

As to the other term from the original theory of Skyrme, it is simply the dot product of the exterior square of the gradient:

$$d\boldsymbol{\phi}^\wedge d\boldsymbol{\phi} = \lambda_2\lambda_3 dy^\wedge dz + \lambda_3\lambda_1 dz^\wedge dx + \lambda_1\lambda_2 dx^\wedge dy. \tag{1.39}$$

The functional from Eq. (1.30) can, therefore, be written in the form

$$E_4(\boldsymbol{\phi}) = \iiint \left(\lambda_2^2\lambda_3^2 + \lambda_3^2\lambda_1^2 + \lambda_1^2\lambda_2^2\right) d^3\boldsymbol{x}. \tag{1.40}$$

Thus d$\boldsymbol{\Phi}$ is defined by a second-order tensor, while d$\boldsymbol{\Phi} \wedge$ d$\boldsymbol{\Phi}$ is defined by a fourth-order tensor. As known, these objects are written with respect to contravariant bases, exactly like the vectors, and in the above functional the Euclidean norm is considered. The Skyrme functional can finally be written as the linear combination (1.33) of the two contributions to the energy, Eqs. (1.38) and (1.40) which gives the energy functional in the form:

$$E_4(\boldsymbol{\phi}) = \iiint \left\{a\left(\lambda_1^2 + \lambda_2^2 + \lambda_3^2\right) + b\left(\lambda_2^2\lambda_3^2 + \lambda_3^2\lambda_1^2 + \lambda_1^2\lambda_2^2\right)\right\} d^3\boldsymbol{x}. \tag{1.41}$$

Details along this line of reasoning can be found in the original work of N. S. Manton [19], as well as in many other works.

A Merit of Manton's Geometrization

The real advantage of Manton's geometrization of Skyrme theory is, in our opinion, that it opens the way for a sound physical basis of reasoning in the realm of nuclear matter. We have thus a way of describing the phenomenology of nuclear matter in well known, even intuitive, terms, descending directly from the roots of Newtonian natural philosophy, the one that established modern theoretical physics.

The whole body of physics going into the stationary principle related functional from Eq. (1.41) is nowadays accumulated mostly into works referring to the large hyper-elastic deformations of rubbers—necessary to modern industry. Here, one usually works under the idea that the functional (1.41) is the energy of deformation of a hyper-elastic material. Generally, the practice shows that the rubber deformation (or that of polymers, in general) proceeds under constant volume, so that the contribution of the variation of volume of matter does not appear in the energy functional. This is implicitly stated by the very form of Manton's functional, should the nuclear matter be described this way.

It is, however, to be noticed that, even under constant volume deformation, from energetical, and therefore physical, point of view the Manton's functional is still a very special one. Indeed, in the area of hyper-elastic deformations, it is only a special instance of the energy of deformation, whereby this one is depicted by the so-called Mooney-Rivlin model for deformation, whose energy density is a linear combination of the two invariants of the deformation

$$I_1 = \lambda_1^2 + \lambda_2^2 + \lambda_3^2;\ I_2 = \lambda_2^2\lambda_3^2 + \lambda_3^2\lambda_1^2 + \lambda_1^2\lambda_2^2. \tag{1.42}$$

Practically, this model describes correctly deformations up to $30 - 40\%$ which, for some rubbers, are even small deformations. However, the Mooney-Rivlin model is the only theoretical model accepting a microscopic theoretical description by means of a Gaussian statistic of the macromolecular chains, so that contemporary theoretical physics almost got stuck within its limits. Another reason for this, more on the practical side we should say, might also be the fact that the range of deformations covered by this description of rubbers is quite enough for application in phenomena of industrial interest. It just happened here what always happens in science when it cannot serve the technology: it is simply discouraged from pursuing its natural path.

However, theoretically speaking, physics cannot afford to say that such a model describes the process of deformation universally. In other words, the variational principle, with such a form of the density of energy, cannot be taken as a fundamental law of nature. So much less, therefore, can we say that such a model will describe the deformations of nuclear matter, where we are not even aware of the amplitude of the deformations. There, in the nucleus, we are in the hazy range of confluence between the matter proper and the space. It is, therefore, to be expected, for instance, that the deformations of nuclear matter are not even reversible in the first place, because of the energy dissipation in the form of particles and heat alike. These are common occurrences in the nuclear physics, and if it is to hold on to the analogy with our inspiring daily nucleus—the Sun—then one can assume that they may be connected with a continuous deformation of the nuclear matter. More than this, the phenomenon of confinement, as understood up to this point in time, when judged in terms of stretches, seems to point out to deformations inside the nucleus amounting to several hundred in percentage.

Then the general idea here would be that the irreversible deformations taking place with dissipation of energy should be described not by the usual invariants of deformation, but by an algebraical combination of them, appearing as an intensity of shearing deformations. This algebraical combination carries, in the theory of deformations, the name of the invariant of von Mises. In terms of the invariants from Eq. (1.42), it can be written as

$$\frac{1}{2}\left(I_1^2 - 3I_2\right) \equiv \left(\lambda_2^2 - \lambda_3^2\right)^2 + \left(\lambda_3^2 - \lambda_1^2\right)^2 + \left(\lambda_1^2 - \lambda_2^2\right)^2. \tag{1.43}$$

Consequently, a "Manton functional" corresponding to such irreversible nuclear deformation should be rather written in the algebraically homogeneous form

$$E(\boldsymbol{\phi}) = \frac{1}{2}\iiint \left(\langle d\boldsymbol{\phi}|d\boldsymbol{\phi}\rangle^2 - 3\langle d\boldsymbol{\phi}\wedge d\boldsymbol{\phi}|d\boldsymbol{\phi}\wedge d\boldsymbol{\phi}\rangle\right)d^3\boldsymbol{x}, \tag{1.44}$$

which is, obviously, out of the limits of Skyrme theory, as it was first conceived. However, this functional suggests a consistent way of description of the nuclear matter, from some other points of view.

The most important of these points of view consists of reducing the Skyrme functional to its first term only, therefore to the usual energy of a harmonic application, however with a nonlinear equation as a result of harmonic principle. This will give directly solitons. R. S. Ward has noticed [22] that in the most general case, of a map from a three-dimensional space to a three-dimensional space, the integrand of the energy functional is actually the trace of the tensor **D** given by

$$D_\alpha^\gamma \equiv g_{\alpha\beta}\frac{\partial\phi^\beta}{\partial x^i}\frac{\partial\phi^\gamma}{\partial x^j}h^{ij}, \tag{1.45}$$

In terms of this, the "pure Skyrme" functional term from Eq. (1.43) is given by

$$I_2 = \frac{1}{2}\left\{(\mathrm{tr}\boldsymbol{D})^2 - \mathrm{tr}\boldsymbol{D}^2\right\}. \tag{1.46}$$

At least in this particular case—we mean for the correspondence from three-dimensional to three-dimensional—we can, therefore, focus on specifying the map $\boldsymbol{\Phi}$ itself, in order to represent the deformation properly. And as the Manton geometrization suggests, we need to generalize it appropriately, in a natural manner so to speak, in order to include the regular hyperbolic geometry as a special case. To understand the geometry and physics of such a generalization, a brief journey into a special theory of deformation of higher orders will be now worth considering.

A Potential of Deformation of Higher Degree

In the theory of deformations, just as in the theory of nuclear matter, the energy functional is not defined by means of criteria physically well-founded, like, for instance, the Lagrangian for a single particle in classical dynamics. Let us elaborate a little on this problem, in order to explain what we mean here by "physically well-founded" criterion.

It is well known that, in the case of a single particle, the Lagrangian represents the difference between the two mechanical energies: kinetic and potential. These appear as the inertial capacity of the particle in motion and as the measurable work of the motion, respectively. The Lagrangian can therefore represent virtually any form of energy that may happen to enter that difference of mechanical energies, and can be actually released in specific conditions, under the law of conservation of energy, of course. It can be released, for instance, in the form of heat which, ever since the times of Joule and Kelvin, is measurable by dissipation phenomena, i.e., mechanically. The classical dynamical equations of motion are then obtained from the Hamilton principle, according to which the true motion of the particle is realized under the condition of stationary action. And the action—i.e.,the time integral of the Lagrangian—can be taken as a measure of the time average of the difference of the two mechanical energies, between the beginning and the end of motion.

Physically, therefore, the Hamilton principle means that along the real space trajectory of the particle motion, any non-mechanical energy release is kept under control: minimized, or at least kept stationary. Energetically, the motion is realized by the transformation of the two mechanical energies into one another, so that the mechanical energy is conserved. Therefore, by this very fact, the Hamilton principle, applied to a particle, is physically a well-founded criterion. As long as we follow the idea of obtaining the pure mechanical and geometrical behavior of a single particle in motion, there is thus a firm physical foundation, expressing this simple fact: when describing the dynamics, eliminate from considerations, or keep under control as much as possible, any non-mechanical energies. If such energies enter the motion of a particle, this motion is "contaminated", so to speak.

In the theory of continuum deformations, as well as in many other physical theories for that matter, the things are however not quite so… physical. The Lagrangian is simply postulated, apparently by considering that in calculations the mechanical work performed by the stresses imposed on matter is already a dissipated energy to be found in different forms in the matter itself. Here the hard part is, for instance, breaking this dissipated energy into its two parts. Indeed, it is known from experience that the deformation has two distinct effects on the structured matter: on one hand, it is the effect of inducing structural changes, on the other hand, it is that of producing heat. Therefore, a part of dissipated energy should be responsible for the deformation-induced structural changes of the matter, the other should be the heat released by internal friction, brought on as a consequence of the deformation process. Thus, for instance, the irreversible deformation of matter occurs along with structural changes leading to the so-called yielding of materials. According to one of the modern views, yielding of matter is a process governed by shear, therefore, to a certain extent, by the average shear strains, in the form given in the equation. And this is what Novozhilov has actually proved [23]: that quantity represents the average shear strain with respect to all possible orientations of the shear planes in a point from an isotropic continuum. Keeping the average process of shearing to a minimum, would then mean the very minimization of the functional from Eq. (1.44). This is one of the first theories of yielding of matter—the so-called von Mises theory [24].

But things get here even more complicated, in the sense that there is no matter of principle involved in the right choice of the measures of deformations and of accompanying tensions. In other words, if the principle does not perform at its full power, part of the blame should be put on the fact that we lack a proper description of the deformation. There are, practically speaking, a bunch of tensor measures of deformation, correlated with some corresponding tensions by means of a certain physical criterion. If this criterion is referring to the mechanical work invested, by means of tensions, into the process of deformation of matter, then a list of theoretically endorsed pairs of tension and deformation measures can be given, conveniently characterizing different deformation processes [20, 25]. That list has, however, the particularity of every other one in this field, namely of describing the deformation ultimately by a gradient. Like, for instance, the previous theory of Manton, which is limited to the classical definition by means of the gradient of the space endomorphism allegedly describing the deformation of the matter. This fact implicitly lays claim to controllability over the deformation of the continuum in all its points, which is obviously an unrealizable task to begin with. Much less, therefore, the description of deformation by stretches can apply at any level of scrutiny into the structure of matter. Indeed, the stretches are defined as experimentally convenient quantities, describing deformation for pieces of matter of definite geometrical shape (for instance, the often quoted, inspirational we should say, rubber bands).

For the matter of indefinite geometrical shape, it is therefore so much proper, when it comes to using the deformation theory as a guide in describing its physics in general, to use also the most general concepts of stress and strain. And the nuclear matter, especially, plainly qualifies for such a treatment. We think therefore proper to use in this respect the most general concepts of stress and strain. And the most general concepts come here naturally with the most general representation of their measures—the 3×3 real matrices—with real eigenvalues. It is these eigenvalues that are supposed to be accessible to our experimental measurements. They are known as principal strains and principal stresses, respectively. This way we still have a limitation—to matrices—but this concept is way richer, allowing us to express more appropriately the mathematical laws describing the matter. Some examples, with particular reference to the definitions related to stretches, will clarify the message we want to convey to our readers. For now, we should just mention the exquisite property of this representation, that of offering a natural realm to the homographic transformation of eigenvalues. This seems to be the basis of a gauge theory generalizing Manton's geometrization.

1.5 A Geometrical Choice of Energy of Deformation

Let us therefore make no other assumption than that the deformations in any point of a continuum are represented by a 3×3 matrix, without being momentarily interested if it is indeed derivable from a gradient or not. It can represent, for instance, a certain variation, at a certain scale, of the metric tensor of the space within which the matter is contained. As an independent property of the space, this tensor should not be always defined by the gradient of the morphism associated with the deformation. Denote by $\mathbf{x}$ this deformation matrix. Its eigenvalues, $\left((\lambda^2 - 1)/2\right.$ in the previous formalism of Manton) are the roots of a third-degree equation—the characteristic equation of that matrix. This is the circumstance allowing us to write a general deformation energy density, based on the mathematical principle of polarity of the binary algebraic forms. This will be explained as we proceed.

Assume, to start with, that we have a binary quadratic form—a homogeneous second-degree polynomial—with coefficients a_0, a_1, and a_2, which happen to have some physical meaning in a certain problem. Assume also that we have a set of cubic binary forms, representing the characteristic polynomials of our deformation matrices which, naturally, evolve in a certain way as the deformation of the continuum proceeds. These cubic polynomials have a joint invariant with our starting quadratic polynomial [26]. The invariant happens to be a quadratic form in the coefficients of the family of cubic polynomials:

$$\frac{1}{9}\psi \equiv a_2\left(X_0X_2 - X_1^2\right) - a_1(X_0X_3 - X_1X_2) + a_0\left(X_1X_3 - X_2^2\right). \tag{1.47}$$

Here X_0, X_1, X_2, X_3 are the coefficients of a generic cubic from our family, written in its binomial (sometimes called canonical) form

$$X_0x^3 + 3X_1x^2 + 3X_2x + X_3. \tag{1.48}$$

X_0—the coefficient of the third-degree term of the polynomial—is responsible for the indetermination in the relations between roots end coefficients, and "x" is the generic nonhomogeneous variable of the cubic. Vanishing of the quadratic form (1.47) implies the apolarity between our starting quadratic form and the Hessian of every member of the family of cubics. As known, the apolarity can be extended to a projective concept, which here comes in handy, inasmuch as the ratios

$$\frac{X_1}{X_0};\ \frac{X_2}{X_0};\ \frac{X_3}{X_0} \tag{1.49}$$

can be taken as nonhomogeneous coordinates of a point in a three-dimensional space of cubics.

From the point of view of the deformation process, let us first notice that the function Ψ from Eq. (1.47) can be taken as a potential that generalizes, in a natural manner so to speak, when it comes to the invariance, the function from Eq. (1.43) to a nonhomogeneous one. Indeed, in cases where de deformations are described by gradients, and are experimentally represented by stretches, the function (1.43) is given by the first term only from the right-hand side of Eq. (1.47). Specifically, the Eq. (1.47) takes there the form

$$\frac{1}{9}\psi \equiv \sum \left(a_2 - 2a_1\lambda_1^2 + a_0\lambda_1^4\right)\left(\lambda_2^2 - \lambda_3^2\right)^2, \tag{1.50}$$

where the summation is understood as being performed over the three positive permutations of the indices of stretches.

More than that, the nonhomogeneous coordinates (1.49) can be taken as realizing a morphism $\boldsymbol{\Phi}$ from the physical three-dimensional space into the three-dimensional space of cubic functions representing the secular equations of the deformation matrices along the process of deformation of the matter. It is therefore to be expected that the solution of a harmonic principle imposed on a form derived from (1.47) in a certain way, will be beneficial in providing a sound representation of the physical and geometrical quantities that properly represent our deformation process. Consequently, it pays to attempt obtaining such a harmonic principle, and this will be done by specifically taking into account the classical Kepler motion as a special case of the principle.

An Illustrative Example of Potential

In order to better understand the geometry involved here and the meaning of the starting quadratic, let us notice, for the moment, that the theory of potential works directly, without the intermediary of stretches—the meaning of which we may have no notion in the nuclear range—as follows. First, if the generic cubic equation of our family is the characteristic equation of the symmetric matrix, say **x**, with the elements $\left(x_{ij}\right)$ say, which happens to represent the deformations, then the matrix **y** having the entries given by the equation

$$y_{ij} = \frac{\partial \psi}{\partial x_{ij}} \tag{1.51}$$

represents the corresponding stresses. Obviously, the converse is also valid. Equation (1.51) suggests that Ψ is a potential indeed, according to the usual rules of working with the concept of potential.

In our conditions, the Eqs. (1.51) are necessarily of the form

$$y_{ij} = \alpha_0\delta_{ij} + \alpha_1 x_{ij} + \alpha_2\left(\boldsymbol{x}^2\right)_{ij} \tag{1.52}$$

because they correlate 3×3 matrices. The problem is to find the coefficients $\alpha_0, \alpha_1, \alpha_2$, which are expected to represent the physical properties of the continuum whose deformation is described by **x**. In order to solve this problem, we need to have the coordinates X_j which, up to a common factor are given by the coefficients of the characteristic equation of **x**:

$$X_0 = 1, X_1 = -\frac{I_1}{3}, X_2 = \frac{I_2}{3}, X_3 = -I_3. \tag{1.53}$$

Here I_1, I_2, I_3 denote the scalar quantities specific to the matrix **x**, which in the case of an Euclidean tensor are its orthogonal invariants. In general, though, they are defined by the formulas

$$\begin{gathered} I_1 = \mathrm{Tr}(x) \equiv t_1, I_2 = \tfrac{1}{2}\left[\left((\mathrm{Tr}(x))^2\right) - \mathrm{Tr}\left(x^2\right)\right] \equiv \tfrac{1}{2}\left(t_1^2 - t_2\right) \\ I_3 = \tfrac{1}{6}\left[2\mathrm{Tr}\left(x^3\right) - 3(\mathrm{Tr}(x))\mathrm{Tr}\left(x^2\right) + (\mathrm{Tr}(x))^3\right] \equiv \tfrac{1}{6}\left(2t_3 - 3t_1t_2 + t_1^3\right). \end{gathered} \tag{1.54}$$

Here we used what seems to be an obvious notation for the traces of powers of the matrix: $t_k = \mathrm{Tr}\left(\mathbf{x}^k\right)$. Using these formulas we have

$$\begin{gathered} X_0X_2 - X_1^2 = \tfrac{1}{18}\left(t_1^2 - 3t_2\right), \\ X_0X_3 - X_1X_2 = -\tfrac{1}{9}\left(t_1^3 - 4t_1t_2 + 3t_3\right), \\ X_1X_3 - X_2^2 = \tfrac{1}{36}\left(4t_1t_3 - 4t_1^2t_2 + t_1^4 - t_2^2\right). \end{gathered} \tag{1.55}$$

Therefore, in order to calculate $\left(y_{ij}\right)$ from (1.51), we need the derivatives of the traces of different powers of the matrix **x**. These are calculated according to the formulas

$$\frac{\partial t_1}{\partial x_{ij}} = \delta_{ij}, \frac{\partial t_2}{\partial x_{ij}} = 2x_{ij}, \frac{\partial t_3}{\partial x_{ij}} = 3\left(x^2\right)_{ij} \tag{1.56}$$

so that, finally, we have for $\alpha_0, \alpha_1, \alpha_2$ the following equations:

$$\begin{gathered} \alpha_0 = a_2t_1 + a_1\left(3t_1^2 - 4t_2\right) + a_0\left(t_1^3 - 2t_1t_2 + t_3\right), \\ \alpha_1 = -3a_2 - 8a_1t_1 - a_0\left(2t_1^2 + t_2\right), \\ \alpha_2 = 3(3a_1 + a_0t_1). \end{gathered} \tag{1.57}$$

On the other hand, the constitutive relation (1.52) is also invertible, i.e., we can write the deformations in terms of stresses in the form

$$\boldsymbol{x} = \beta_0\boldsymbol{e} + \beta_1\boldsymbol{y} + \beta_2\boldsymbol{y}^2, \tag{1.58}$$

where **e** is the identity 3×3 matrix. Here β_0, β_1, and β_2 are connected with α_0, α_1, and α_2 by the system of equations

$$\begin{gathered} \beta_0 + \beta_1\alpha_0 + \beta_2\left(\alpha_0^2 + \alpha_2^2I_1I_3 + 2\alpha_1\alpha_2I_3\right) = 0, \\ \alpha_1\beta_1 + \left[2\alpha_0\alpha_1 - 2\alpha_1\alpha_2I_2 + \left(I_1^2 - I_2\right)\alpha_2^2\right]\beta_2 = 1, \\ \alpha_2\beta_1 + \left[\alpha_1^2 + 2\alpha_1\alpha_2I_1 + 2\alpha_0\alpha_2 + (I_3 - I_1I_2)\alpha_2^2\right]\beta_2 = 0. \end{gathered} \tag{1.59}$$

This shows that there are nontrivial states of stress, corresponding to null deformations, as in the case of vacuum, or classical ether. These nontrivial states are determined exclusively by the starting binary quadratic form, through the equations

$$\begin{aligned} \beta_0 + \beta_1\alpha_0 + \beta_2\alpha_0^2 &= 0, \\ \alpha_1\beta_1 + 2\alpha_0\alpha_1\beta_2 &= 1, \\ \alpha_2\beta_1 + \left(\alpha_1^2 + 2\alpha_0\alpha_2\right)\beta_2 &= 0, \end{aligned} \tag{1.60}$$

where, according to (1.47), we take

$$\alpha_0 = 0;\ \alpha_1 = -3a_2;\ \alpha_2 = 9a_1. \tag{1.61}$$

Thus, we have

$$\beta_0 = 0;\ \beta_1 = -\frac{1}{3a_2};\ \beta_2 = \frac{a_1}{3a_2^3}. \tag{1.62}$$

Our starting quadratic form should therefore physically represent such limit states of matter. Generally, however, the constitutive relations (1.52) and (1.49) cannot be simultaneously inverted, because there are separate deformations that do not involve stresses, as well as stress phenomena that do not involve deformations. The matter for which these phenomena are simultaneous should therefore have special properties.

The relation (1.52) shows that the stresses can all vanish for certain deformation processes, still related to the starting quadratic form of this theory. Indeed, the system of equations

$$\alpha_0 = \alpha_1 = \alpha_2 = 0 \tag{1.63}$$

has a nontrivial solution for the invariants of deformations. This solution is given by the following formulas for the traces of powers of the deformation matrix as functions of the coefficients of the starting binary quadratic

$$t_1 = -3\frac{a_1}{a_0};\ t_2 = 3\frac{2a_1^2 - a_0a_2}{a_0^2};\ t_3 = 3\frac{4a_1^3 - 3a_0a_1a_1}{a_0^3}. \tag{1.64}$$

In cases where the starting quadratic form has real roots, say $(u \pm v)$, these equations give

$$t_1 = 3u;\ t_2 = \left(u^2 + v^2\right);\ t_3 = 3u\left(u^2 + 3v^2\right). \tag{1.65}$$

Using now the formulas (1.53) and (1.54), we find easily the coefficients of the characteristic equation of the corresponding matrix in the form

$$\frac{c_0}{1} = \frac{c_1}{-u} = \frac{c_2}{u^2 - \frac{v^2}{2}} = \frac{c_3}{-u\left(u^2 - \frac{3v^2}{2}\right)}. \tag{1.66}$$

This cubic is the center of the quadric given in Eq. (1.47), and has a remarkable property. Having the roots given by

$$u - \sqrt{\frac{3}{2}}v, u, u + \sqrt{\frac{3}{2}}v \tag{1.67}$$

one can see that these roots are strictly determined by its Hessian. Consequently, this is a cubic having the phase zero. Therefore, the center of the quadric (1.47) is not a reference only from a projective point of view, but also algebraically. Continuing along this line, the Hessian of this reference cubic is given by the equations

$$\frac{A_0}{1} = \frac{A_1}{-2u} = \frac{A_2}{u^2 + \frac{v^2}{2}} = -\frac{v^2}{2}. \tag{1.68}$$

Obviously, this is a polar conjugate of our starting binary quadratic. There is also a family of cubics conjugated with (1.67) in the sense that their Hessians are apolar with this one. This is a hyperboloid, having the equation

$$\left(u^2 + \frac{v^2}{2}\right)(X_0X_2 - X_1^2) + u(X_0X_3 - X_1X_2) + (X_1X_3 - X_2^2) = 0. \tag{1.69}$$

The center of this quadric is given by

$$\frac{b_0}{1} = \frac{b_1}{-u} = \frac{b_2}{u^2 + \frac{v^2}{2}} = \frac{b_3}{-u\left(u^2 + \frac{3v^2}{2}\right)} \tag{1.70}$$

and it is a cubic having only one real root, i.e., its roots are

$$u - i\sqrt{\frac{3}{2}}v, u, u + i\sqrt{\frac{3}{2}}v. \tag{1.71}$$

Obviously, we have here a duality between the two cases.

The bottom line is that the starting quadratic form is something that, at least in the case of deformations as we agree to describe them here, can be determined without ambiguity. Thus, for instance, there is a state of strain in space, specifically in the space of the center of force associated with a Kepler orbit, and determined through its geometrical characteristics. Starting from this state of strain, we can find a family of states of deformation, also represented by Kepler motions, but otherwise undecided. In the macroworld, such phenomenon can be documented indeed: the Sun and the planets are in a continuous deformation, and the Moon for instance, slowly but continuously departs from Earth. This very phenomenology should therefore apply, by the classical analogy, to the atomic and nuclear model.

References

1. Mittag, L., & Stephen, M. J. (1992). Conformal transformations and the application of complex variables in mechanics and quantum mechanics. *American Journal of Physics, 60*, 207–211.
2. Cayley, A. (1859). A sixth memoir upon quantics. *Philosophical Transactions of the Royal Society of London, 149*, 61–90. Reprinted in The collected Mathematical Works, Vol. II, p. 561, Cambridge University Press, 1889.
3. Klein, F. (1897). Sur la Géométrie Dite Non Euclidienne, Annales de la Faculté des Sciences de Toulouse, Série 1, Tome 11, pp. G1–G62; (French translation of the original German memoir from Mathematische Annalen, Vol. IV, pp. 573–625, 1871).
4. Eells, J., & Sampson, J. H. (1964). Harmonic mappings of Riemannian manifolds. *American Journal of Mathematics, 86*, 109160.
5. Misner, C. W. (1978). Harmonic maps as models for physical theories. *Physical Review D, 18*, 4510–4524.
6. Ernst, F. J. (1968). New formulation of the axially symmetric gravitational field problem I. *Physical Review, 167*, 1175–1178; New formulation of the axially symmetric gravitational field problem II, *Physical Review, 168*, 1415–1417.
7. Ernst, F. J. (1971). Exterior algebraic derivation of Einstein field equations employing a generalized basis. *Journal of Mathematical Physics, 12*, 2395–2397.
8. Feynman, R. P. (1948). Space-time approach to non-relativistic quantum mechanics. *Reviews of Modern Physics, 20*, 367–387.
9. Feynman, R. P. (1949). The theory of positrons. *The Physical Review, 76*, 749–759.
10. Fock, V. A. (1964). *The theory of space, time and gravitation*. New York: Macmillan Company.
11. Evrard, G. (1995). Minimal information in velocity space. *Physics Letters A, 201*, 95–102.
12. Atiyah, M. F., Sutcliffe, P. (2001). The geometry of point particles, arXiv:hep-th/0105.1794v2
13. Atiyah, M. F., & Sutcliffe, P. (2004). Skyrmions, Instantons, Mass and Curvature, arXiv:hep-th/0411052v1
14. Skyrme, T. H. R. (1988). The origins of skyrmions. *International Journal of Modern Physics A, 3*, 2745–2751.
15. Skyrme, T. H. R. (1994). Selected Papers, with Comentary, edited by G. E. Brown, World Scientific Series in 20th Century Physics, Vol. 3
16. Slobodeanu, R. (2009). *On the geometrized skyrme and faddeev models*, arXiv:0809.4864v5 [math.DG].
17. Skyrme, T. H. R. (1961). A non-linear field theory. *Proceedings of the Royal Society of London, A260*, 127–138.
18. Rogers, C., & Schief, W. K. (2002). *Bäcklund and darboux transformations*. Cambridge, UK: Cambridge University Press.
19. Manton, N. S. (1987). Geometry of skyrmions. *Communications in Mathematical Physics, 111*, 469–478.
20. Hill, R. (1968). On constitutive inequalities for simple materials I. *Journal of the Mechanics and Physics of Solids, 16*, 229–248.
21. Ogden, R. W. (1972). Large deformation isotropic elasticity: On the correlation of theory and experiment for incompressible rubberlike solids. In Proceedings of the Royal Society of London, Vol. A326, pp. 565–584, *Large Deformation Isotropic Elasticity: On the Correlation of Theory and Experiment for Compressible Rubberlike Solids*, Proceedings of the Royal Society of London, Vol. A328, pp. 567–583.
22. Ward, R. S. (2004). Skyrmions and Faddeev-Hopf Solitons. *Physical Review D, 70*, 061701(R).
23. Novozhilov, V. V. (1952). On the physical meaning of the invariants of tensions used in the theory of plasticity (О Физическом Смысле Инвариантов Напряжения Используемых в Теории Пластичности). *Prikladnaya Matematika i Mekhanika, 16*, 617–619.
24. von Mises, R. (1913). Mechanik der festen Körper im plastisch-deformablen Zustand, Nachrichten von der Gesellschaft der Wissenschaften zu Göttingen, Mathematisch-Physikalische Klasse. *No., 1*, 582–592.

25. Hill, R. (1978). Aspects of invariance in solid mechanics. *Advances in Applied Mechanics, 18*, 1–76.
26. Burnside, W. S., & Panton, A. W. (1960). *The theory of equations*. Dover Publications.

Chapter 2
From Special to General Relativity Through Complex Potential

2.1 The Special Relativity as Absolute Geometry

There is no way of representing a physical point in a space but rather by physical quantities or by kinematical or dynamical quantities describing that point. Once this representation is accomplished, the remaining problem is purely theoretical: the connection of these quantities with positions in the space of our intuition. First, the points in the four-dimensional manifold of spacetime events are represented by the coordinates of the positions of uniform motions starting from the origin and the radius of the sphere of propagation with the center in the origin. Within the framework of space kinematics, the limitation of possible velocities of inertial matter epitomizes the mathematical idea of confinement, as presented in Chap. 1. From the very same point of view, the matter in a planetary model shaped by a Kepler motion can be thought of as properly confined, provided we assume spatial extension of the material components of this model. Thus, the difference between the two models of mathematical confinement—the relativistic and the Keplerian models—restrains upon the sole fact that, while relativistic, the limit velocity is a constant, in the case of Kepler motion analysis, the limit velocity is variable with the Hertz material particle from the matter in revolution, as well as with the material particle from the matter thought to create the central field. Both cases, however—the relativity and the Kepler problem—are about a velocity space. This calls for a general presentation of the mathematical confinement of the velocities, whereby the limit velocity is variable, to be able to describe the most general case. The equation of a Kepler ellipse in polar coordinates is

$$\frac{\dot{a}}{x} = \frac{k}{\dot{a}} = v_1 \cos \phi + v_2 \sin \phi \tag{2.1}$$

M. Agop and S. A. Irimiciuc, *Multifractal Theory of Motion*,
https://doi.org/10.1007/978-981-97-6410-5_2

is indeed an equation for velocities. To wit, it connects the velocity $\left(a'/x\right)$ to the magnitude of the actual position vector **x** of the mobile material point along its orbit. This material point is identified by its initial velocity (v_1, v_2), which is chosen to describe that orbit. One can say that a Kepler orbit is a trajectory corresponding to the given initial velocity, which is a closed curve—an ellipse—as Kepler inferred for the first time from the analysis of the data of Tycho Brahe for the planet Mars—if, and only if, this initial velocity satisfies the following inequality:

$$\left(\frac{k}{a}\right)^2 - v_1^2 - v_2^2 > 0 \tag{2.2}$$

As we already noted, this condition describes a particular relativistic ensemble of particles, out of all possible Hertz material particles from the ensemble, giving the interpretation of the kernel, or nucleus, of an electron in the planetary model. Indeed, if the electron has a closed orbit—assuming the condition of an ideal one—in the sense that the shape is the same after traversing the orbit—then, from a classical point of view, all of the constituent material particles from the interpretative ensemble of the nucleus of that electron should have closed orbits in the interior of a torus in space. The ideal condition also assumes that the Hertzian material particles have trajectories all along the Kepler orbit of the electron.

Now, if we denote by X a point in this space of velocities, then a coordinate representation is given by a quadruple of (until further notice, real) numbers:

$$X \equiv (c, \mathbf{v}) \tag{2.3}$$

The norm of these points is then given by the quadratic form:

$$(X, X) \equiv c^2 - v^2. \tag{2.4}$$

Thus, the points that satisfy the condition $(X, X) = 0$ are geometrically shaping an absolute for this geometry. Physically, they represent the propagation of light if the limit velocity c is a constant. Specifically, points with positive norms represent inertial motions, while points with negative norms represent, for instance, de Broglie waves in the regular case of special relativity or other ensembles of Hertz material particles in the case of Kepler motion.

The norm (2.4) induces an internal multiplication of points via the known "polarization" procedure

$$(X_1, X_2) \equiv c_1 c_2 - \mathbf{v}_1 \cdot \mathbf{v}_2. \tag{2.5}$$

Here, an obvious correspondence is understood between the indices of points and the indices of coordinates. This internal product helps in describing a straight line in space, which is the essential concept necessary in constructing a metric, at least from the differential geometric point of view. The straight line joining two points X_1

and X_2 is the locus of the points

$$X = \lambda X_1 + \mu X_2 \tag{2.6}$$

with λ and μ variable numbers representing homogeneous parameters of points along the line. This straight line intersects the absolute line at two points, with the homogeneous parameters partially established by the quadratic equation:

$$(X, X) \equiv \lambda^2(X_1, X_1) + 2\lambda\mu(X_1, X_2) + \mu^2(X_2, X_2) = 0 \tag{2.7}$$

From this equation, we can determine only the ratios of these two parameters—the nonhomogeneous coordinates along the line as its two roots, viz.

$$t \equiv \frac{\lambda}{\mu} = -\frac{(X_1, X_2)}{(X_1, X_1)} \pm \frac{\sqrt{(X_1, X_2)^2 - (X_1, X_1)(X_2, X_2)}}{(X_1, X_1)} \tag{2.8}$$

However, these ratios are sufficient for our purpose of building a metric of space. Indeed, a metric represents the distance between two infinitesimally close points; therefore, we need to determine the distance between points. The quantity that reduces to the distance between two positions in the Euclidean case turns out to be the cross-ratio of four points on a straight line, two of which are fixed and used as a reference frame of that straight line. More precisely, the distance is, up to a numerical factor, the logarithm of this cross-ratio, the so-called Laguerre formula [for a brief pertinent presentation, see, for instance, [1]]. Given two points $X_{1,2}$, the straight line joining them contains all points of the form $X = tX_1 + X_2$, where t is a nonhomogeneous parameter. To define the distance between two points, we arbitrarily choose a reference frame formed by two other points on their straight line, $X_{3,4}$, and define the cross-ratio of the four points. As the cross-ratio of points is simply the cross-ratio of the corresponding nonhomogeneous parameters t, we have, for instance:

$$(X_1 \cdot X_2; X_3, X_4) = \frac{t_1 - t_3}{t_1 - t_4} \frac{t_2 - t_3}{t_2 - t_4} \tag{2.9}$$

According to Laguerre's formula, the distance is simply proportional to the logarithm of this quantity. It depends, of course, on the reference frame—the second pair of points on the straight line—but this ambiguity can be substantially reduced if we refer the construction to the absolute of space and take the reference frame on the line as given by the two points of the straight line located on the absolute. First, parameter t has the values. $t_2 = 0$ for point X_2 and $t_1 = \infty$ for point X_1. In view of this, the cross-ratio (2.9) takes the simple form of a usual ratio:

$$(X_1 \cdot X_2; X_3, X_4) = \frac{t_4}{t_3} \tag{2.10}$$

Incidentally, we need to note that the choice of the reference frame on the line, as given by two points on the absolute, has the advantage of allowing standardization of the construction of the metric, as every pair of points in space has a corresponding pair of points on the absolute: the points where the corresponding straight line intersects the absolute, as we said. The parameters $t_{3,4}$ are then given by the two ratios from Eq. (2.8), so Eq. (2.10) becomes

$$(X_1 \cdot X_2; X_3, X_4) = \frac{(X_1, X_2) + \sqrt{(X_1, X_2)^2 - (X_1, X_1)(X_2, X_2)}}{(X_1, X_2) - \sqrt{(X_1, X_2)^2 - (X_1, X_1)(X_2, X_2)}}. \tag{2.11}$$

With this, we can construct a differential version of the distance, which is the metric of space. Assuming that the two points X_1 and X_2 are infinitesimally close, $X_1 = X, X_2 = X + dX$, we can calculate the necessary quantities in Eq. (2.11) in mixed finite and differential forms:

$$\begin{aligned} &(X_1, X_2) = (X, X) + (X, dX), \\ &(X_2, X_2) = (X, X) + 2(X, dX) + (dX, dX), \\ &(X_1, X_2)^2 - (X_1, X_1)(X_2, X_2) = (X, dX)^2 - (X, X)(dX, dX). \end{aligned} \tag{2.12}$$

Now, at least in the real range, we can accept that the quantity $(X, dX)/(X, X)$ is infrafinite of the first order in the sense of GeorgescuRoegen, while also forcing the condition that in the arithmetic of this first order, the square root of $(dX, dX)/(X, X)$ is such an infrafinite. Thus, the cross-ratio (2.12) can be expanded in a series, and to the first infinitesimal order, it is

$$(X_1, X_2, X_3, X_4) = 1 + 2\sqrt{\left(\frac{(X, dX)^2}{(X, X)}\right) - \frac{(dX, dX)}{(X, X)}} \tag{2.13}$$

Furthermore, the logarithm of this quantity is, to the same infinitesimal order, the second term from the right-hand side of this equation so that the absolute metric can be written, up to an arbitrary constant factor, in the form:

$$(ds)^2 = \left(\frac{(X, dX)}{(X, X)}\right)^2 - \frac{(dX, dX)}{(X, X)} \tag{2.14}$$

This expression is also valid under larger spatial conditions, such as complex points, the definition of the absolute as a general quadric in this space, or even as a general quantic. Thus, we have a general form of the absolute metric valid for any type of quadric, independent of the dimension of space and the nature of its elements. Therefore, let us return to our specific case, as given by Eq. (2.4). If the coordinates of the points are real, then the absolute quadric is a two-sheet hyperboloid, as we know it from analytic geometry. This description can also be used for the case of complex coordinates. However, in that instance, the geometrical image is not as simple as in the real case: there, we have to do with the intersection of two real quadrics and therefore with a real conic in space. Regardless of these details, we refer to a direct calculation of the metric given by Eq. (2.14). This will be based on the identification $dX \equiv (dc, d\mathbf{v})$, which gives the following quadratic form as a metric based on this representation:

$$(ds)^2 = \frac{c^2(d\mathbf{v})^2 - (\mathbf{v} \times d\mathbf{v})^2}{\left(c^2 - \mathbf{v}^2\right)^2} + \frac{\mathbf{v}^2(dc)^2 - 2c(\mathbf{v} \cdot d\mathbf{v})(dc)}{\left(c^2 - \mathbf{v}^2\right)^2}. \tag{2.15}$$

This metric can be related to some known and physically important situations. Therefore, let us consider the equation obtained in greater detail.

We wrote here the absolute metric as a sum of two terms for a particular reason: in the framework of special relativity, the first term from equation (2.15) is the metric of velocity space and is a direct consequence of the law of composition of relativistic velocities. In this taking, one can say that the whole metric (2.15), itself, generalizes the relativistic metric of the velocity space, which can be obtained from (2.15), for instance, in cases where the first component of the point X is a constant: the differential of a constant is always zero, and the second term in (2.15) vanishes. In other words, there is no infrafinite four-vector velocity in relativistic physics, as we inherited it from Maxwell electrodynamics. However, such a metric also characterizes a classical Kepler orbit, as in Eq. (2.1), a description that can be obtained by identifying the vector with the initial condition of motion and then taking . Every closed Kepler orbit is thus described by some initial velocity with a magnitude smaller than this value, as in Eq. (2.2) above. The metric (2.15) could therefore be applied for describing the structure of matter contained in the nucleus of the revolving body of the planetary model of physics, but there is a drawback. Indeed, in this interpretation, the matter of the revolving material point can be taken as a "swarm" of Hertz material particles, as described by Joseph Larmor [2]. Then, the value of c is variable with each particle of the material point in revolution. Moreover, as each particle is uniquely labeled by its initial conditions, there should be a correlation between the limit velocity c and the label of the particle. The nature of such a correlation can be inferred directly from (2.15) because this formula still discloses another condition in which the metric reduces to the usual "relativistic" one, in addition to maintaining a constant c. This condition is given by the differential equation

$$vdc - 2cdv = 0 \therefore v^2/c = \text{ const}, \tag{2.16}$$

which guarantees the vanishing of the second term in (2.15). In other words, in our present generalization of SR, there are two cases in which the metric becomes purely relativistic. One of these is the usual case of constancy of the value of limit velocity, which precludes the infrafinite differential measure for the time component in the realm of velocities taken as four-vectors: there are only regular finite four-vectors and infrafinite three-vectors there. On the other hand, there is a differential measure in this range, given by Eq. (2.16), involving the connection between the time component and space components of the four vectors. These last velocities are "reciprocal" with respect to the "sphere" given by the initial conditions of the Kepler problem. Such a reciprocity is akin to a "de Broglie duality", as it was between the phase and wave-group velocities. However, this time, it is the inertial matter that imposes it upon light. In general, therefore, one can assume that is not a constant, regardless of this last situation, and describe a universal and special relativity starting from a "double" relativistic metric to be obtained from (2.15) in two special cases: one for considering the light in ether and the other considering the matter in ether. For the first case, we cannot decide anything yet, but the last case can always be described, indeed, by two absolute metrics corresponding to two limit velocities.

To obtain the gist of the method, let us assume the case of constant: only after analyzing this case can we properly improve upon it. The absolute metric (2.15) is

$$(ds)^2 = \left[1 - \left(\frac{v}{c}\right)^2\right]^{-2} \left[\left(\frac{dv}{c}\right)^2 - \left(\frac{v}{c} \times \frac{dv}{c}\right)^2\right] \tag{2.17}$$

In the three-dimensional velocity space of relativity, this metric can be written in the form

$$(ds)^2 = \left(\frac{d\boldsymbol{q}}{1-\boldsymbol{q}^2}\right)^2 - \left(\frac{\boldsymbol{q} \times d\boldsymbol{q}}{1-\boldsymbol{q}^2}\right)^2; \boldsymbol{q} \equiv \frac{v}{c} \tag{2.18}$$

while in the two-dimensional case [3], it can be written in the form

$$(ds)^2 = \left(\frac{d\boldsymbol{e}}{1-\boldsymbol{e}^2}\right)^2 - \left(\frac{\boldsymbol{e} \times d\boldsymbol{e}}{1-\boldsymbol{e}^2}\right)^2; \boldsymbol{e} \equiv \frac{v}{c}, c \equiv \frac{k}{a'} \tag{2.19}$$

Here, e is precisely the eccentricity vector of the orbit. There is no formal difference between the two formulas, (2.18) and (2.19), except for the dimension of the velocity space, and the latter is controlled by gravitation according to Newtonian precepts. Keeping this last condition in reserve, (2.19) can be obtained from (2.18) just by choosing to work in one of the planes of coordinates of the velocity three-dimensional space.

To settle our ideas, we choose to work on formula (2.19), with the Cartesian components of the eccentricity vector, for which it is more obvious how the space extension of matter can enter our reasoning. The eccentricity vector e represents the relative position of the center of orbit in the dynamical Kepler problem with respect to the center of force. Therefore, it can be taken as a rough measure of the space extension of the matter generating the force field. One can surely assume that the geometry of the space containing the matter that generates the force field in the classical Kepler problem is not an Euclidean geometry but a Lobachevsky geometry. In other words, we have a geometry of the space region containing forces: it is the hyperbolic geometry. To make this statement even more obvious, the absolute metric (2.19) can be cast into the Beltrami-Poincaré form

$$(ds)^2 = -4\frac{dhdh^*}{(h-h^*)^2} \equiv \frac{(du)^2+(dv)^2}{v^2} \tag{2.20}$$

by the transformation

$$h = u + iv \equiv \frac{e_2 + i\sqrt{1-e^2}}{1-e_1}; h^* \equiv u - iv. \tag{2.21}$$

There is an element of arbitrariness left with this representation, leading to the necessity of our reverse interpretation, which is the objective of the present chapter of this work. Indeed, the metric is conformal to a Euclidean metric, and this latter metric is known to be invariant with respect to rotations in the plane. We need to point out how this arbitrariness appears in hyperbolic geometry.

In (u, v) coordinates, this geometry exhibits a special invariance connected with the Bäcklund transformation. Following a known routine, we take a point P in the hyperbolic plane as described in the coordinates (u, v) so that its vector variation can be written as [4]

$$\boldsymbol{dP} = \begin{pmatrix} du \\ dv \end{pmatrix} \equiv \sigma^1 \boldsymbol{e}_1 + \sigma^2 \boldsymbol{e}_2 \boldsymbol{e}_1 = \begin{pmatrix} v \\ 0 \end{pmatrix}, \boldsymbol{e}_2 = \begin{pmatrix} 0 \\ v \end{pmatrix}. \tag{2.22}$$

This formula clearly shows that the hyperbolic plane is conformal to the Euclidean plane. If we take the two differential forms

$$\sigma^1 = \frac{du}{v}, \sigma^2 = \frac{dv}{v}.$$

As a coframe coframe, the Beltrami-Poincaré (2.20) is the sum of their squares. Thus, the hyperbolic plane can be treated as a constant curvature surface, where the parallel transport of a vector, $\boldsymbol{V}$, is given by the following equations:

$$dv^\alpha + v^\nu \Omega_\nu^\alpha = 0, \tag{2.23}$$

where v^α are the components of the vector V in a local reference frame, and Ω is a Frenet-Serret matrix of differential elements. Now, along the paths on which the vector V is transported by parallelism, there is only a normal component of the variation of the vector, which can be written as

$$dV = \left(v^\alpha \Omega_\alpha^3\right) e_3{}' \therefore e_3{}'dV = v^\alpha \Omega_\alpha^3. \tag{2.24}$$

Here, the normal vector to the hyperbolic plane is taken as the unit length to simplify the argument. Applying this theory to the displacement vector from Eq. (2.22), we have that, if

$$d\sigma^{\alpha} + \sigma^{\nu}\Omega_{\nu}^{\alpha} = 0,$$

then the vector dP is parallel-transported in the hyperbolic plane. The paths along which this transport is realized are then geodesics of the plane [5]. Their equations can therefore be written in a purely differential form, independent of any continuity parameter:

$$d\frac{du}{v} - \frac{du}{v}\frac{dv}{v} = 0, d\frac{dv}{v} + \frac{du}{v}\frac{du}{v} = 0. \tag{2.25}$$

This form of the equation involves only symmetric differentials, i.e. differentials in the sense of Leibniz—not exterior differentials—and we use it here to draw attention to the fact that the equations can even be fractal: there is no need to refer them to a continuity parameter. Once this reference is made, however, the differentials become "fluxions" in the plain sense of Newton's initial theory, representing the convention of the definition of "finite" with the associated "infrafinite" and "transfinite" in the acceptance of Nicholas Georgescu-Roegen. Now, by proper identification, Eq. (2.25) makes the connection differential form (du/v) conspicuous. By expanding the equation by performing the following differentiations, we have

$$\frac{d^2u}{v} - 2\frac{dudv}{v^2} = 0, \frac{d^2v}{v} + \frac{(du)^2 - (dv)^2}{v^2} = 0.$$

The geodesic equations can be obtained by using the metric (2.20) as a Lagrangian and then solving the appropriate associated variational problem. One can then calculate the components of d^2P along the geodesics, and with them, the second fundamental form of the surface locally represents the hyperbolic plane. According to Eq. (2.24), the second fundamental form of this surface coincides with its first fundamental form, i.e. with the Beltrami–Poincaré metric itself:

$$\acute{e}_3 \cdot d^2P = -\left\{\left(\sigma^1\right)^2 + \left(\sigma^2\right)^2\right\}. \tag{2.26}$$

The unit vector of the Beltrami–Poincaré metric [6] is therefore given by a conformity, starting from the unit vector of the Euclidean plane:

$$v = (\cos\phi)e_1 + (\sin\phi)e_2;\ v^1 = \cos\phi,\ v^2 = \sin\phi. \tag{2.27}$$

According to (2.23) and (2.25), this vector is parallel-transported in the hyperbolic plane if and only if

$$d\phi + \frac{du}{v} = 0; \omega_2^1 = -\frac{du}{v}.$$

Now, since the Beltrami–Poincaré metric is itself conform–Euclidean, it does not change in a rotation of the coframe, which comes to a Bäcklund transformation of the differentials

$$\sigma^1 = \cos\phi\frac{du}{v} + \sin\phi\frac{dv}{v}; \sigma^2 = -\sin\phi\frac{du}{v} + \cos\phi\frac{dv}{v} \tag{2.28}$$

However, something still changes, and this is the connection form. Indeed, by differentiation here, we have

$$d\sigma^1 = (d\phi)\sigma^2 + \cos\phi d\frac{du}{v} + \sin\phi d\frac{dv}{v},$$
$$d\sigma^2 = -(d\phi)\sigma^1 - \sin\phi d\frac{du}{v} + \cos\phi d\frac{dv}{v}.$$

Therefore, along the geodesics given by equation (2.28), we must have

$$d\sigma^1 = \left(d\phi + \frac{du}{v}\right)\sigma^2; d\sigma^2 = -\left(d\phi + \frac{du}{v}\right)\sigma^1. \tag{2.29}$$

Thus, the newly rotated coframe (2.28) is still parallel-transported along the geodesics, only with a new connection form, which represents "an update", as it was, of the original one by the differential of the rotation angle. This is the meaning of the Bäcklund transformation (2.28).

However, there is more to it: as it turns out, the system of the three differential forms [7]

$$\Omega^1 = d\phi + \frac{du}{v}, \Omega^2 = \cos\phi\frac{du}{v} + \sin\phi\frac{dv}{v},$$

$$\Omega^3 = \sin\phi\frac{du}{v} + \cos\phi\frac{dv}{v}$$

formed, as one can see, with the new connection form and the new coframe (2.28), is an algebraically closed system in the sense that they constitute a coframe for an sl(2, *R*) type algebra. This fact is not as obvious, at "the first sight", as it was, but it can be made obvious by an exterior differentiation of the indicated 1-forms, for the Maurer-Cartan relations here turn out to be of the characteristic form:

$$d \wedge \Omega^1 = \Omega^2 \wedge \Omega^3; d \wedge \Omega^2 = -\Omega^3 \wedge \Omega^1; d \wedge \Omega^3 = -\Omega^1 \wedge \Omega^2 \tag{2.30}$$

In fact, the Killing–Cartan quadratic form that describes this algebra as a Riemannian space is that from equation

$$ds^2 = \left(\Omega^1\right)^2 - \left(\Omega^2\right)^2 - \left(\Omega^3\right)^2 = (d\phi)^2 + \frac{2}{v}(d\phi)(du) - \frac{dv^2}{v^2}$$

a typical three-dimensional quadratic Lorentz–Minkowski form. It describes a Riemannian space that reduces to the hyperbolic plane in cases where ϕ is itself the "angle of parallelism", i.e., a connection. This space is the central feature of our physics here, as it is representative of matter per se. Let us elaborate on this aspect of the problem because it is designed to illuminate the kinds of forces that are described by Riemannian geometry and what is their relationship with the manifold of events, i.e., with spacetime.

The transformation (2.21) clarifies what physical kind of geometrical beings are represented by the complex variables h and h^*: they are gauging lengths proper or parameters related to these lengths, such as the semiaxes of the Kepler orbits and their eccentricity or, in the case of light, the semiaxes of the "ellipsoid of elasticities" of ether. If we take this last case as typical for gauging a continuum, then the concepts are algebraically settled by the theory of cubic equations, describing the semiaxes of an ellipsoid, which are always expressed by the roots of a cubic equation as eigenvalues of a tensor. In this case, the geometry is no longer an Euclidean one but should be submitted to some rules of gauging. These rules are obvious if we express them in terms of the two basic tools of the theoretical physics of the universe: the Wien-Lummer hohlraum and the Einstein elevator. By definition, these tools ask that the laws governing the universe must be the same regardless of the size of the universe and invariant for the transition between different sizes. In other words, they should be scale invariant.

2.2 General Relativity and Ernst Potential

In space–time theory [8, 9], the metric tensor g_{ik} is determined by the space–time energy tensor by means of Einstein's field equations, which, after convenient adjustment of the Einsteinian gravitational constant and metric signature (system of physical units $G = c \equiv 1$ where G is Newton's constant and c is the light vacuum speed), are written as

$$G_{ik} = -6\pi T_{ik},\ 1, k = 0, 1, 2, 3, \tag{2.31}$$

where T_{ik} is the energy tensor and G_{ik} is Einstein's tensor, written in terms of the contracted curvature tensor R_{ik} and curvature invariant R as [8]:

$$G_{ik} = R_{ik} - \frac{1}{2}Rg_{ik}. \tag{2.32}$$

The postulation of Eq. (2.31) to determine the metric tensor has, as a consequence, the fact that their solutions are nonunique, not even in the simplest cases.

This is why one usually speaks about "classes of solutions" of Einstein's equations and, correspondingly, about "classes of metrics". These classes are characterized by a priori conditions: flatness at infinity, symmetries, etc. Even in the simplest cases, such as vacuum or stationary electromagnetic vacuum, there are a multitude of solutions, with various known metrics as particular cases. Nevertheless, in many cases, it appears possible of their rational 'classification', meaning that they can be obtained one from the other by means of certain groups of transformation.

To provide evidence, we used the method of "complex potentials", or "Ernst potentials", whose mechanism is particularly well described by Israel and Wilson [10]. In this respect, let us write the space–time metric in the form

$$ds^2 = f\left(dx^0 + \omega_\alpha dx^\alpha\right)^2 - f^{-1}\gamma_{\alpha\beta}dx^\alpha dx^\beta, \tag{2.33}$$

where the summation over repeated indices within a term is understood, and we denote it as follows:

$$\begin{aligned}\gamma_{\alpha\beta} &= g_{0\alpha}g_{0\beta} - g_{00}g_{\alpha\beta},\\ \omega_\alpha &= g_{0\alpha}/g_{00},\\ f &= g_{00}.\end{aligned} \tag{2.34}$$

Here, $\left(\gamma_{\alpha\beta}\right)$ is a three-dimensional metric that can be used for raising and lowering indices. The three-vector ω_α is defined only up to a gradient, corresponding to the arbitrary choice of the origin of the temporal coordinate x^0. Therefore, it can be used to obtain a vector invariant with respect to the arbitrary choice of the origin by means of the "curl" operation in the metric $\gamma_{\alpha\beta}$, which is

$$\frac{1}{f^2}\tau = -curl\ \omega \tag{2.35}$$

or, explicitly

$$\frac{1}{f^2}\tau^\alpha = -\frac{1}{\sqrt{\gamma}}\varepsilon^{\alpha\beta\mu}\partial^\beta\omega^\mu, \tag{2.36}$$

where $\gamma = \det\left(\gamma_{\alpha\beta}\right)$, and $\varepsilon_{\alpha\beta\mu}$ is the three-dimensional totally antisymmetric symbol. Now, suppose that in the space-time of metric (2.33), there exists an electromagnetic stationary field derived from the potential A_k:

$$F_{ik} = \partial_k A_i - \partial_i A_k. \tag{2.37}$$

By virtue of time independence, the electric components of the electromagnetic field tensor are

$$F_{0\alpha} = \partial_\alpha A_0, \tag{2.38}$$

while Maxwell's equations are written as

$$\partial_\alpha\left\{(-g)^{1/2}F^{i\alpha}\right\} = 0, \tag{2.39}$$

where $g = \det(g_{ik})$ can be written as

$$(-g)^{1/2} = f^{-1}\gamma^{1/2}. \tag{2.40}$$

Now, we show that the full set of Maxwell's equations can be reduced to a single equation with partial derivatives for a scalar complex field. To this end, one observes that (2.39), with $i = \beta$, makes it possible to write $F^{\beta\alpha}$ in terms of a "magnetic" potential ϕ as follows:

$$F^{\beta\alpha} = f\gamma^{-1/2}\varepsilon^{\beta\alpha\mu}\partial_\mu\phi. \tag{2.41}$$

The remaining components can then be expressed by means of (2.38) and (2.41), according to

$$F^{\alpha 0} = \omega_\beta F^{\beta\alpha} + F_{0\beta}\gamma^{\beta\alpha}. \tag{2.42}$$

Considering (2.41), (2.38), and (2.36), Eq. (2.9) for becomes

$$\mathrm{div}\left(f^{-1}\nabla A_0\right) = -f^{-2}\boldsymbol{\tau}\cdot\nabla\phi. \tag{2.43}$$

Using cyclic identity:

$$\varepsilon^{\alpha\beta\mu}\partial_\mu F_{\alpha\beta} = 0,$$

where is given by (2.41); we still have

$$div\left(f^{-1}\nabla\phi\right) = f^{-2}\tau\cdot\nabla A_0 \tag{2.44}$$

We mention, again, that all the differential operators are expressed in terms of the three-dimensional metric $\gamma_{\alpha\beta}$.

Next, let us define the complex potential:

$$\psi = A_0 + i\phi, i = \sqrt{-1}. \tag{2.45}$$

By means of (2.43)–(2.45), we then have

$$\mathrm{div}\left(f^{-1}\nabla\psi\right) = f^{-2}\tau\cdot\nabla\psi, \tag{2.46}$$

expressing the whole set of Maxwell's equations.

Analogously, one can express the Ricci tensor R_{ik} by means of the vector $\boldsymbol{G}$ defined as

$$2fG = \nabla f + i\tau. \tag{2.47}$$

The components of R_{ik} are then defined by

$$-f^{-2}R_{00} = \text{div }\mathbf{G} + \mathbf{G}\cdot\left(\mathbf{G}^* - \mathbf{G}\right) \tag{2.48}$$

$$-2if^{-2}R_0^\alpha = \gamma^{-1/2}\varepsilon^{\alpha\beta\mu}\left(\partial_\mu G_\beta + G_\beta G_\mu^*\right) \tag{2.49}$$

$$f^{-2}\left(\gamma_{\mu\alpha}\gamma_{\nu\beta}R^{\alpha\beta} - \gamma_{\mu\nu}R_{00}\right) = R_{\mu\nu}(\gamma) + G_\mu G_\nu^* + G_\mu^* G_\nu \tag{2.50}$$

Here, $R_{\mu\nu}(\gamma)$ denotes the Ricci tensor of the three-dimensional metric $\left(\gamma_{\alpha\beta}\right)$, and by an asterisk, (*) is marked as the complex conjugate. The energy tensor of the electromagnetic field is then given by the following general formula:

$$-4\pi T_{ij} = g^{lm}F_{im}F_{jl} - \frac{1}{4}g_{ij}F_{lm}F^{lm}.$$

According to our previous results

$$\frac{1}{2}F_{ik}F^{ik} = (\nabla\phi)^2 - (\nabla A_0)^2$$

We still have

$$8\pi f^{-1}T_{00} = (\nabla\phi)^2 + (\nabla A_0)^2 \tag{2.51}$$

$$\begin{aligned} 4\pi f^{-1}T_0^\alpha &= \gamma^{-1/2}\varepsilon^{\alpha\mu\nu}\left(\partial_\mu\phi\right)\left(\partial_\nu A_0\right), \\ -4\pi f^{-1}T^{\alpha\beta} &= \left(\partial^\alpha\phi\right)\left(\partial^\beta\phi\right) + \left(\partial^\alpha A_0\right)\left(\partial^\beta A_0\right) \end{aligned} \tag{2.52}$$

$$-\frac{1}{2}\gamma^{\alpha\beta}\left[(\nabla\phi)^2 + (\nabla A_0)^2\right] \tag{2.53}$$

Since the energy tensor of the electromagnetic field is a null trace tensor [9], Einstein's Eq. (2.31) can be written in a simpler form as

$$R_{ik} = -8\pi T_{ik}.$$

Combining these equations with (2.19) and (2.22), we have

$$curl\ \tau = -4(\nabla\phi\times\nabla A_0) = i\ curl\left(\psi\nabla\psi^* - \psi^*\nabla\psi\right),$$

so that the vector quantity is

$$\tau + i(\psi^* \nabla \psi - \psi \nabla \psi^*)$$

derived from a potential, e.g. φ, defined up to an arbitrary additive constant

$$\tau + i(\psi^* \nabla \psi - \psi \nabla \psi^*) = \nabla \varphi. \tag{2.54}$$

Now, we define a new complex potential

$$E = f - \psi \psi^* + i\varphi \tag{2.55}$$

such that (2.54) and (2.47) yield

$$f\boldsymbol{G} = -\frac{1}{2}\nabla E + \psi^* \nabla \psi. \tag{2.56}$$

Substituting (2.56) in (2.48) and (2.51), by means of (2.46), we still have

$$f\nabla^2 E = \nabla E(\nabla E + 2\psi^* \nabla \psi). \tag{2.57}$$

We also observe that (2.46) can be written in a form analogous to (2.57), that is

$$f\nabla^2 \psi = \nabla \psi(\nabla E + 2\psi^* \nabla \psi). \tag{2.58}$$

By means of the complex potentials E and ψ and the metric $\gamma_{\alpha\beta}$, one can express everything. According to (2.55), we have

$$f = \frac{1}{2}(E + E^*) + \psi^* \psi; \tag{2.59}$$

while (2.50) and (2.53) yield

$$\begin{aligned} -f^2 R_{\alpha\beta}(\gamma) &= \frac{1}{2}(E, E^*)_{\alpha\beta} + \psi(E, \psi^*)_{\alpha\beta} \\ &+ \psi^*(E^*, \psi)_{\alpha\beta} - (E + E^*)(\psi, \psi^*)_{\alpha\beta}, \end{aligned} \tag{2.60}$$

where we denoted

$$2(A, B)_{\alpha\beta} \equiv (\partial_\alpha A)(\partial_\beta B) + (\partial_\beta A)(\partial_\alpha B).$$

Therefore, the complete system of equations of gravitational and electromagnetic fields in vacuum is now expressed by (2.57), (2.58), and (2.60).

The idea of using a complex potential to express the equations of the gravitational field belongs to F.J. Ernst, first concerning the axially symmetric metric [11, 12]and then for the general case of a stationary metric [13]. This last case of Ernst's investigation concerns the gravitational field in a vacuum and is composed of the following equations:

$$\frac{1}{2}(\varepsilon+\varepsilon^*)\nabla^2\varepsilon=\nabla\varepsilon\cdot\nabla\varepsilon, \tag{2.61}$$

and

$$-(\varepsilon+\varepsilon^*)R_{\alpha\beta}(\gamma)=(\partial_\alpha\varepsilon)(\partial_\beta\varepsilon^*)+(\partial_\beta\varepsilon)(\partial_\alpha\varepsilon^*), \tag{2.62}$$

which are obtained from (2.57), (2.58), and (2.60) by setting $\psi=0$ and $\varepsilon=f+i\varphi$ in (2.55).

The importance of this approach to the solution of Einstein's equations is revealed by the fact that Eqs. (2.57), (2.58) and (2.60) (or (2.61) and (2.62) for the purely gravitational case) can be inverted. Namely, $(\gamma_{\alpha\beta})$ can be any three-dimensional metric that is conveniently chosen, for which (2.60) and (2.62) represent the curvature tensor. In this way, our attention is focused on solving (2.57) and (2.58), or Ernst's Eq. (2.61), which has been intensively investigated by several authors (see, e.g. [14, 15]).

The general procedure for constructing the metric (2.33) is to obtain f and ω from the solutions of (2.57), (2.58), and (2.60), followed by A_0 and ϕ, and, finally, the components of the electromagnetic field. The most difficult choice proves to be the metric $(\gamma_{\alpha\beta})$: it has to possess a certain liberty to be compatible with Eq. (2.32) for the purely gravitational field. Otherwise, the use of Eqs. (2.60) and (2.62) is not essential: there is a theorem concerning the generation of families of solutions, starting from their existence for the purely gravitational field [14].

Next, we provide evidence of a certain procedure for solution generation, starting with already-known solutions, as shown by Kinnersley [15]. We shall follow his investigation but restrict ourselves only to the bidimensional case. To this end, let us introduce two complex potentials, u and v, in (2.61), instead of the complex potential ε, by means of

$$\varepsilon=\frac{u-v}{u+v}. \tag{2.63}$$

Obviously, one of the potentials u and v is redundant, and their ratio is the only thing that counts. In other words, one can use the potentials u and v so that definition (2.63) is compatible with Ernst's Eq. (2.61). By substituting (2.63) into (2.61), we find the following set of symmetric differential equations as sufficient conditions:

$$\begin{aligned}(uu^*-vv^*)\nabla^2u&=2(u^*\nabla u-v^*\nabla v)\nabla u,\\ (uu^*-vv^*)\nabla^2v&=2(u^*\nabla u-v^*\nabla v)\nabla v.\end{aligned} \tag{2.64}$$

This definition allows us to introduce a complex bidimensional space of vector components u and v

$$z=(u,v)$$

metrized by the indefinite quadratic form with a metric matrix:

$$\eta_{\alpha\beta} = diag(1, -1), \alpha, \beta = 1, 2. \tag{2.65}$$

In this case, Eq. (2.64) can be written in a condensed form as

$$z^{\alpha} z^{*}_{\alpha} \nabla^2 z^{\beta} = 2 z^{*}_{\alpha} \nabla z^{\alpha} \nabla z^{\beta}.$$

This is a vector equation whose group of covariants leaves the invariant metric (2.65), which invariates the Hermitian scalar products. Therefore, we have to do with the unitary matrix group 2×2:

$$\begin{bmatrix} a & b \\ -b^* & a^* \end{bmatrix} \tag{2.66}$$

subject to condition

$$aa^* + bb^* = 1.$$

We mention, nevertheless, that by virtue of (2.63), this group acts on Ernst's potential as a homographic group: denoting $\varepsilon = (\xi - 1)/(\xi + 1)$, then the complex homography

$$\xi' = \frac{a\xi - b^*}{b\xi + a^*} \tag{2.67}$$

shall correspond to the transformation given by matrix (2.66). Therefore, matrix (2.66) is defined only up to an arbitrary normalization factor.

The property of being invariant with respect to the homography of Ernst's equation can also be made straightforward, but the method used above provides important evidence: the Ernst potential can be considered an inhomogeneous coordinate for a certain binary domain [16]. This fact should subsequently be explained by means of some operational significance given to this coordinate system.

The existence of theorems regarding the generation of solutions of the field equations with the help of already-known solutions [14] allows us to restrict our investigation to Ernst's Eq. (2.61), which is of maximum generality. Ernst [11, 12] mentioned two essential points regarding these equations. First, it is invariant under complex conjugation: the complex conjugate of ε satisfies the same equations and, therefore, can play the role of the complex gravitational potential. Second, Ernst demonstrated that the formal Eq. (2.63) can be derived from a variational principle. This last aspect has not been extensively studied in the literature, even if, in our opinion, it is of maximum formal importance. Indeed, Eq. (2.63) is one of the Euler–Lagrange equations associated with the functional

$$2\int \frac{\nabla\varepsilon\nabla\varepsilon^*}{(\varepsilon+\varepsilon^*)^2}\gamma^{1/2}d^3x \tag{2.68}$$

the other one is its complex conjugate. Here, we mention, once again, that the dot product is understood in the "ambient" metric $(\gamma_{\alpha\beta})$.

The method of obtaining Eq. (2.61) did not simultaneously reveal its complex conjugate, and its validity was expressed as an a posteriori observation. However, the variational principle attached to functional (2.68) does not make any difference between them.

2.3 Metrics of the Lobachevsky Plan and Solutions of the Gravitational Field

In view of these observations, the problem of the validity of the variational principle itself appears since the existence of the complex conjugate equation exceeds the frame within which it has been deduced. Regarding Einstein's equations, only one of these two variants can be used.

The answer to this problem can be found at the end of the previous paragraph. The equation is invariant with respect to the group of homographies—a continuous group with three essential parameters [16]. Without loss of generality, we consider these homographies to be real; this is always possible and can be performed by applying a Cayley-type [17] transformation to matrix (2.66) to make it real. In addition, we use a complex variable h instead of a variable ξ and denote by $\overleftarrow{h}$ the complex conjugate of h.

Therefore, consider a real homographic transformation of the variable h:

$$\begin{aligned} h' &= \frac{ah+b}{ch+d},\; a,b,c,d \in \Re, \\ ad &- bc \neq 0. \end{aligned} \tag{2.69}$$

The conjugates of the two complex quantities h and h' are connected, obviously, by the same transformation:

$$\overline{h'} = \frac{a\overleftarrow{h}+b}{c\overleftarrow{h}+d} \tag{2.70}$$

Transformations (2.69) and (2.70) form a group of two variables with three parameters (one of the quantities a, b, c, d is not essential), as it can easily be verified. This group leaves invariant the differential quadratic form:

$$-4\frac{dhd\overleftarrow{h}}{(h-\overleftarrow{h})^2} \tag{2.71}$$

In view of (2.69) and (2.70), we have

$$dh' = \frac{(ad-bc)dh}{(ch+d)^2}$$

$$d\overline{h'} = \frac{(ad-bc)d\overleftarrow{h}}{(c\overleftarrow{h}+d)^2}$$

$$h'-\overline{h'} = \frac{(ad-bc)\left(h-\overleftarrow{h}\right)}{(ch+d)\left(c\overleftarrow{h}+d\right)}$$

which immediately yields

$$\frac{dh'd\overline{h'}}{\left(h'-\overline{h'}\right)^2} = \frac{dhd\overleftarrow{h}}{(h-\overleftarrow{h})^2}.$$

The quadratic form (2.71) is a well-known metric, namely, the metric of the Lobachevsky plane, in Poincaré's representation [18]. It can be written in real form, which is most commonly encountered, by $h = u + iv$. The result is

$$-4\frac{dhd\overleftarrow{h}}{(h-\overleftarrow{h})^2} = \frac{du^2+dv^2}{v^2} \tag{2.72}$$

Recently, the properties of this metric have been emphasized in connection with some theories on supergravity [19]. We could be interested in its behavior as long as h represents an Ernst potential, in which case the group (2.69) and (2.70) would continuously generate solutions. Now, we have to answer the following question: is there some equation with all these solutions? Such an equation is necessary to obtain at least one of the concrete forms of h, the other ones being then generated through the medium of (2.69).

Supposing that the "ambient" "metric" is $\gamma_{\alpha\beta}$, the Lagrangian function corresponding to metric (2.42), according to the usual procedure [20], is written as follows:

$$L = \gamma^{\alpha\beta}\frac{\frac{\partial h}{\partial x^\alpha}\frac{\partial \overleftarrow{h}}{\partial x^\beta}}{(h-\overleftarrow{h})^2} = \frac{\nabla h\nabla\overleftarrow{h}}{(h-\overleftarrow{h})^2}$$

The variational principle

$$\delta \int L\gamma^{1/2} d^3x \tag{2.73}$$

leads then to the following equations:

$$\begin{aligned} \left(h - \overleftarrow{h}\right)\nabla^2 h &= 2\nabla h \nabla h \\ \left(h - \overleftarrow{h}\right)\nabla^2 \overleftarrow{h} &= 2\nabla \overleftarrow{h} \nabla \overleftarrow{h} \end{aligned} \tag{2.74}$$

For $h = i\varepsilon$, we are left with Ernst's equations.

The fact that the field equations are equivalent to a variational principle has been put into evidence by Matzner and Misner [21] in connection with an axially symmetric field. These authors did not arrive at the complex potential, but their method can be applied in our case as follows. Let us suppose that the field is described by the variables $\left(Y^j\right)$, for which we have discovered the metric

$$h_{ij} dY^i dY^j \tag{2.75}$$

in an ambient space of metric

$$\gamma_{\alpha\beta} dx^\alpha dx^\beta \tag{2.76}$$

In this situation, the field equations are derived from a variational principle connected to the Lagrangian:

$$L = \gamma^{\alpha\beta} h_{ij} \frac{\partial Y^i}{\partial x^\alpha} \frac{\partial Y^j}{\partial x^\beta} \tag{2.77}$$

In our case, the metric (2.75) is given by (2.71), where the field variables are h and $\overleftarrow{h}$, or equivalently, the real and imaginary parts of h.

Therefore, if this variational principle is accepted as a starting point, the main purpose of gravitational field research would be to produce metrics of the Lobachevsky plane (or related to them), which are closely connected to Einstein's field equations, by means of the formalism described in the previous paragraph. However, as one can see, such a variational principle exceeds the theoretical frame of employing the Ricci tensor. In fact, one can give examples of its applicability for cases that have nothing to do with Einstein's field equations.

Indeed, the metric of the Lobachevsky plane can be produced as a Cayleyan metric of an Euclidean plane, for which the absoluteness is a circle with a unit radius [18]. In this way, the Lobachevsky plane can be put into biunivocal correspondence with the interior side of this circle. The general procedure of metrization of a Cayleyan space starts with the definition of the metric as an anharmonic ratio, which is extensively described in many classical volumes (see, e.g. [22]), so that we skip the details and give only the final results, which are necessary for our following investigation. Let us suppose that the absoluteness of the space is represented by the quadratic form $\Omega(X, Y)$, where X denotes any vector. The Cayleyan metric is then given by the differential quadratic form

$$\frac{-ds^2}{k^2} = \frac{\Omega(dX, dX)}{\Omega(X, X)} - \frac{\Omega^2(X, dX)}{\Omega^2(X, X)} \tag{2.78}$$

where $\Omega(X, Y)$ is the duplication of $\Omega(X, X)$ and k is a constant connected to the space curvature.

In the case of the Lobachevsky plane, we have

$$\begin{aligned} \Omega(X, X) &= 1 - x^2 - y^2, \\ \Omega(X, dX) &= -xdx - ydy, \\ \Omega(dX, dX) &= -dx^2 - dy^2, \end{aligned} \tag{2.79}$$

which yields

$$\frac{ds^2}{k^2} = \frac{\left(1 - y^2\right)dx^2 + 2xydxdy + \left(1 - x^2\right)dy^2}{\left(1 - x^2 - y^2\right)^2} \tag{2.80}$$

Now, we perform the coordinate transformation

$$x = \frac{h\overleftarrow{h} - 1}{h\overleftarrow{h} + 1}, \; y = \frac{h + \overleftarrow{h}}{h\overleftarrow{h} + 1}, \tag{2.81}$$

the metric (2.80) becomes identical to (2.71)

$$\frac{ds^2}{k^2} = -4\frac{dhd\overleftarrow{h}}{(h - \overleftarrow{h})^2} \tag{2.82}$$

The absoluteness $1 - x^2 - y^2 = 0$ becomes a straight line $\Im h = 0$, while straight lines of the Euclidean plane go to circles with centers on the real axis of the complex plane (h).

2.4 Rindler-Type Motion

The previous theory is applicable to any case of the existence of a real conic, as it is easy to demonstrate that the Cayleyan metric attached to this conic is the metric generated by the transformation group, which leaves it invariant. Before studying the metric (2.82) in more detail, let us give an example of the potential applicability of the variational principle of Matzner–Misner regarding this metric. This example has nothing to do with the curvature tensor because it is based on the idea of avoiding the use of this tensor.

In 1960, Rindler [23] showed that motions performed in general space–time on torsion-free trajectories with constant curvature represent a natural generalization of the so-called hyperbolic motions (with constant acceleration) of the Minkowskian space–time. These motions are particularly important since they are described in coordinates analogous to Kruskal coordinates [24] and, in addition, by means of these coordinates, were put into evidence—for the first time—of the nonequivalence of the second quantization in Minkowskian frames, on the one hand, and uniformly accelerated frames, on the other hand [25]. An interesting result of this theory is the fact that an observer engaged in a uniformly accelerated motion in the Minkowskian vacuum is situated in a thermal bath, and its temperature is directly proportional to his or her acceleration [26]. For the one-dimensional case, the equation of Minkowskian hyperbolic motion is written as [24]:

$$Z^2 - c^2T^2 = \frac{c^4}{\alpha^2}, \tag{2.83}$$

where α is the magnitude of acceleration, and the other quantities possess the usual significance. The last equation explains the term "hyperbolic" associated with this motion. The quantity

$$X = \frac{c^2}{\alpha} \tag{2.84}$$

is called the "Rindler coordinate", which is attached to the hyperbolic motion. Equation (2.83) then yields

$$Z^2 = c^2T^2 + X^2, \tag{2.85}$$

giving us the possibility of considering the rest condition $Z =$ const. for a particle engaged in a hyperbolic motion. Equation (2.85) can be interpreted as absoluteness in coordinates:

$$x = \frac{X}{Z}, y = \frac{cT}{Z}, \tag{2.86}$$

and obtain the metric (2.82) by taking

$$h = \frac{y \pm i\sqrt{1 - x^2 - y^2}}{1 - x}, \tag{2.87}$$

or, in view of (2.86)

$$h = \frac{cT \pm i\sqrt{z^2 - c^2T^2 - X^2}}{Z - X}. \tag{2.88}$$

The corresponding Ernst equations, written in the form (2.74) with , are equivalent to the following pair of real equations:

$$\begin{aligned} v\nabla^2 u - 2\nabla u \nabla v &= 0, \\ v\nabla^2 v - (\nabla v)^2 + (\nabla u)^2 &= 0, \end{aligned} \tag{2.89}$$

where the differential operators are taken in an arbitrary three-dimensional metric. implies , where

$$v = \frac{\sqrt{Z^2 - X^2}}{Z - X}, \tag{2.90}$$

meaning that the Ernst's potential is purely real. In this case, the field is characterized by the second Eq. (2.89), which is easily solved. If the change of variable is performed, it reduces to the Laplace equation

$$\nabla^2 \xi = 0. \tag{2.91}$$

Let us solve this equation for one of the simplest cases, i.e., the spherical symmetry of . In spherical coordinates , Eq. (2.91) is written as

$$\frac{1}{r^2}\frac{d}{dr}\left(r^2 \frac{d\xi}{dr}\right) = 0$$

with solution

$$\xi(r) = C_1 - \frac{C_2}{r} \tag{2.92}$$

or, by the use of (2.90) and an adequate choice of the integration constants:

$$\frac{Z + X}{Z - X} = e^{-\frac{K}{r}}. \tag{2.93}$$

Equation (2.93) can be used to extract the coordinates :

$$X = -Z \tanh\left(\frac{K}{r}\right). \tag{2.94}$$

In the limit , one can write

$$X = -\frac{ZK}{r}$$

or, by means of (2.84):

$$\alpha = -\kappa^2 r. \tag{2.95}$$

with $\kappa^2 = ZK/c^2$. It then follows that $T = 0$, α can be considered an acceleration in a radial oscillatory motion or, more correctly—if passing to the limit leading to (2.95)—is considered a centripetal acceleration in a uniform circular motion. How can we interpret this fact? The variational principle used to obtain the above result emerges from the metric (2.82), which is invariant with respect to a certain group of coordinate rational transformations. If we adopt the equivalence principle [27], then for an arbitrary point, the Rindler coordinate X represents the gravitational field intensity, while the mentioned group represents the transition between various gravitational fields acting on that point. The simultaneous action of these gravitational fields is a motion of rotation characterized by the centripetal force given by (2.95) at the initial (Minkowski!) moment far from the common center of the field-generating sources.

Therefore, by replacing the principle of independence of simultaneous actions, as a linear composition of field intensities at a point, with the a priori invariance of field action with respect to a certain group, one could conceive a theory of gravitation free of the usual inherent contradictions of actual theory [28]. As we have mentioned, the starting point of our previous discussion was not the curvature tensor but rather a certain (hyperbolic) motion. However, if similar results obtained by means of Ernst's equation, that is, the Matzner–Misner variational principle, prove their strength, eliminating some contradictions in the actual theory of gravitation, this means that this variational principle is of maximum generality and leads to results that can involve space curvature [29].

Expressing our belief in the last possibility, next, we provide a detailed description of the metric (2.53), followed by its generalization and, implicitly, generalization of the variational principle generated by it, as well as an attempt at an operational definition of the fields constructing the Lagrangian.

References

1. Duval, C., & Guieu, L. (2000). The Virasoro group and lorentzian surfaces: the hyperboloid of one sheet. *Journal of Geometry and Physics, 33*, 103–127; arXiv: mathDG/9806135v1.
2. Larmor, J. W. (1900). On the statistical dynamics of gas theory as illustrated by meteor swarms and optical rays. *Nature, 63*(1624), 168–169; British Association Report, September 1900, pp. 632–634.

3. Mazilu, N., Agop, M., & Merches, I. (2019). *The mathematical principles of scale relativity physics: The concept of interpretation*. CRC Press, Boca Raton.
4. Flanders, H. (1989). *Differential forms, with applications to the physical sciences*. Dover Publications.
5. Struik, D. J. (1988). *Lectures on classical differential geometry*. Dover NY.
6. Shirokov, A. P. (1988). On the geometry of tangent bundles of homogeneous spaces. *The Works of the Geometry Seminar of the Kazan University, 18*, 153–163. (in Russian).
7. Agop, M., & Merches, I. (2019). *Operational procedures describing physical systems*. C.R.C Press.
8. Ionescu-Pallas, N. (1980). *General relativity and cosmology, (in Romanian)*. Scientific and Encyclopedic Publishing House.
9. Adler R., Bazin M., & Schiffer M. (1972). *Introduction to general relativity*. Mc Graw Hill.
10. Israel, W., & Wilson, G. A. (1972). *Journal of Mathematics and Physics, 13*, 865.
11. Ernst, F. J. (1968). *Physical Review, 167*, 1175.
12. Ernst, F. J. (1968). *Physical Review, 168*, 1415.
13. Ernst, F. J. (1971). *Journal of Mathematics and Physics, 12*, 239.
14. Ray J. R., & Wei M. S. (1977). *Nuovo Cimento, B, 42*, 151.
15. Kramer D. et al. (E. Schmutzer, Editor) (1980). *Exact solutions of Einsteins field equations*. Cambridge Univ. Press.
16. Barbilian, D. (1967). Opera Mathematica. *Romanian Academy Publishing House, Bucharest, I*, 286.
17. Fr´e R. (1979). *Acta Phys. Polonica, B*, 10, 383.
18. Efimov, N. (1981). *G´eometrie Sup´erieure*. Mir.
19. Vysotsky, M. I. (1986). *Iader. Fiz., 43*, 1338.
20. Fock, V. A. (1964). *The theory of space*. Macmillan.
21. Matzner, R. A., & Misner, C. W. (1967). *Physical Review, 154*, 1229.
22. Barbilian, D. (1971). *Opera Didactica*, Technical Publishing House, Bucharest, Vol. *I, II, III*(1968), 1974.
23. Rindler, W. (1960). *Physical Review, 119*, 2082.
24. Rindler, W. (1966). *Amer. J. Phys., 34*, 1174.
25. Fulling S. A. (1973). *Phys. Rev., D*, 7, 2850.
26. Unruh W. G. (1976). *Phys. Rev., D*, 14, 870.
27. Einstein, A. (2004). *Relativity: The special and the general theory*. Barnes and Noble Library of Essential Reading.
28. Popescu, I. N. (1982). *Gravitation (in Romanian)*. Scientific and Encyclopedic Publishing House.
29. Fock, V. A. (1959). *The theory of space*. Pergamon Press, New York.

Chapter 3
Geometrical and Physical Frames

3.1 Considerations of the Frame Concept

Both experimental and theoretical physics pay tribute to the notion of frame. The position and/or motion of a body can only be described with respect to a certain frame. From the experimental point of view, the frame is given by a system of bodies—or, in a more abstract manner—of points relatively fixed with respect to each other than intermediate measurements of distances and directions. Interference of the speculative thinking then made it possible to detach the notion of coordinates of a point in a given frame: spatially, the point is localized by three real numbers—its coordinates—and, if it moves, also by the time moment determined by a horologe attached to the point when its localization is determined.

Theoretically speaking, the events or phenomena are described in coordinates chosen as conveniently as possible. Usually, the coordinate systems are chosen to satisfy a certain type of symmetry or to offer the possibility of direct, experimental testing via theoretical investigation.

In principle, calculations can be performed in any coordinate system since there are many transformation formulas from one system to another. This reciprocal transformability led to detachment of the notion of a frame from its initial, experimental meaning. A frame usually used in geometry is the frame of the vector's tangent to the coordinate lines (the so-called natural frame [1]). Because transitions between such frames are given as derivatives of some well-known functions, these transformations are called holonomic. Compared to these purely mathematical frames, the physical frame is marked by a high degree of arbitrarily: not always its abstract defining elements—the basic vectors—can define tangent to them coordinate lines. As a result, transformations between two such frames cannot be expressed by usual functions, and the basis of such frames is called un-holonomic [2]. Nevertheless, these frames are commonly used in experimental physics, the holonomic frames being only abstractions necessary for calculations, as we shall hereinafter show.

M. Agop and S. A. Irimiciuc, *Multifractal Theory of Motion*,
https://doi.org/10.1007/978-981-97-6410-5_3

From a physical point of view, a coordinate system must correspond to reality, being measurable or interpreted in terms of measurable quantities. However, in almost all cases, the coordinates are not measurable. In astronomy, for example, the position of a star can be determined by either Cartesian coordinates or spherical coordinates, such as the azimuth, height, and distance to the star. In both cases, we have to do with quantities whose significance or, alternatively, the effectively determined value are unknown. For example, the Cartesian coordinates cannot be measured, even if the space is supposed to be absolutely Euclidean. In astronomy, one usually employs the second set of coordinates, but the radial distance—obtained by physical means—is quantitatively affected by a large scale of arbitrariness. Even qualitatively, if the space is non-Euclidean, the distance changes its initial significance, while the angles do not indicate the real direction but only an apparent, local one. Regardless, these frames—or parts of them, which could eventually play the role of a "theodolite"—maintain their significance, which ignores displacements and takes into account only rotations.

These facts are important since, among all notions used in physics, the notion of direction is somewhat transcendental. For example, the experimental device must obey the classical laws with good precision. In other words, the difference between classical laws (applicable to measurement instruments) and quantum laws (concerning micro-objects), implying the scale of measurement of mass, length, and time, extends on quantities implying displacements but not on directions, implying rotations only. There are two aspects here we must point out. The first concerns the intuitive fact that a direction—regardless of how it is settled—can serve as a frame for both microscopic and macroscopic phenomena. The second refers to the fact that the uncertainty relations, implying angular variables and associated operators, demand precautions in their definition [3]; this could display a certain "classicism" of the considered variables.

The indiscernibility of the coordinate systems, together with the fact that in measurements, only "anholonomic" physical frames are used, clearly shows the necessity of reporting calculations to the latter ones. In addition, only these frames are used, for example, in the operational definition of space–time metric elements. However, as observed by Ionescu Pallas [4], even models of the universe defined by ingenious operational procedures, such as Milne's kinematic model, can be reformulated by means of the notion of metrics. There exists a firm basis for metric theories: Einstein's general relativity. What exactly is missing in physics (or, more limited, in cosmology) to overcome the actual status quo is the physical interpretation—i.e., in a physical picture—of the coordinates defining the metric. In other words, if a system of coordinates can usually define a frame (of the tangent vectors), the actual problem is inverse: given a frame, find the compatible coordinates. The previous theory of complex potential (see Chap. 2) allows us to come close to solving this problem since it becomes clear that the only "holonomic" coordinates belong to the "ambient" three-dimensional metric, which can be a working hypothesis.

Therefore, these coordinates are connected to the measured quantities (e.g., a potential) only by a certain function as a solution of a partial differential equation. However, this fact implies the extension of the operational procedures beyond the usual ones: the measurement of distances, even by light signals, and the synchronization of horologes in the same way. These procedures are intrinsically connected, and since synchronization is marked by a high degree of arbitrarily, the definition of distance has the same property.

3.2 Frame Through Directions in Space

One way of defining a measurement procedure connected to the physical frame, as previously accepted, consists of speculation based on quantum measurability, that is, on Hermitian operators whose per values (eigenvalues) stand for measured values of the quantities with attached operators. To explain such a point of view, we first have to accept that on a frame can be settled directions, indifferently from our working phenomenological limits (micro- or macrocosms). Specifically, we admit that—artificially (with rational intervention) or intrinsically—a physical frame is endowed with "theodolite devices" whose orientation is frame-independent and, most of all, independent of the way the frame is considered (e.g., independent of the orientation of the three attached Cartesian axes). Regarding the theodolite device, we have to realize that inside this structure, some interactions occur due to the fields caught from the vision direction; by means of these interactions, some quantities connected to this field are experimentally determined. Due to its nature (construction), device measurements do not depend on the direction of observation or parameters associated with this direction. In other words, the internal process of measurement is independent of direction.

Under these circumstances, a theodolite must be characterized by a Hermitic operator that can depend on orientation but has eigenvalues independent of direction. Such an operator can be constructed in the case of two real and distinct eigenvalues. For example, if these eigenvalues are ± 1, the operator can be expressed as [5]

$$Q = \begin{vmatrix} \cos\theta & \sin\theta e^{i\varphi} \\ \sin\theta e^{-i\varphi} & -\cos\theta \end{vmatrix} \tag{3.1}$$

where θ and φ are the azimuth and height of the direction, or, better to say, its parameters on the unit sphere. The two eigenvectors of the matrix Q are defined up to an arbitrary phase factor, but the ratios of their components, which are the fixed points of the matrix Q, are considered as homography:

$$z_1 = \cot\frac{\theta}{2}e^{i\varphi}, z_2 = -\tan\frac{\theta}{2}e^{i\varphi} \tag{3.2}$$

are in biunivocal correspondence with the considered direction. These parameters (z_1, more frequently) serve to express direction as a point function on the sphere of unit radius, considering this sphere as a one-sheet hyperboloid of imaginary generatrixes [6]. The matrix

$$M = xE + yQ, \tag{3.3}$$

where x and y are real and E is the 2×2 unit matrix, which has the same fixed points as Q but with the following eigenvalues:

$$\lambda_1 = x + y, \lambda_2 = x - y, \tag{3.4}$$

which is direction independent. Consequently, the matrix (3.3), which is obviously Hermitic, satisfies our demands to characterize a tool, and we shall use it in the investigation to follow. The physical significance of the eigenvalues λ_1 and λ_2 or, better, of x and y remains to be determined.

Parameters x and y have a certain "internal" freedom, which must be considered. Indeed, the results of measurements in some direction reflect a "sum" of influences of the same nature from all spatial directions that do not explicitly interfere. The mentioned "sum" (for example, superposition of gravitational and/or electromagnetic fields) depends on the physical structure of the theodolite and could be expressed as a certain internal freedom concerning the measurement process if we consider it—for example—as being given by a linear group of transformations in x and y.

It is very difficult, obviously, to a priori put into evidence such a group, but we will try to do so by analogy with some known cases. To this end, let us consider the similitude transformations of matrix M:

$$M' = L^{-1}MR, \tag{3.5}$$

where L and R are two 2×2 matrices. Obviously, such transformations correlate matrices M and M', each of which has its own characteristics for different directions and different eigenvalues. Since these matrices affect directions, they are not of much help, but—as we shall see—they can be used for some analogy.

3.3 Directions in Space at Large Scale

As we noted in the previous chapter, obtaining a Cayleyan metric could be useful in many cases due to its connection with the Matzner–Misner variational principle. Since x and y can be considered measured field variables, we repeat our attempt to obtain such an invariant, but this time with respect to transformations (3.5). In this case, the absoluteness is given by the matrices M of the null determinant [6], which signifies that one or both eigenvalues are zero. By virtue of previous observations, such matrices characterize the device, while no "measurement", in no direction, is performed, which is the cosmic background. The procedure of the metric construction was given in [1, 6], with the following result

$$-\frac{ds^2}{k^2} = \frac{dx^2 - dy^2}{x^2 - y^2} - \frac{(xdx - ydy)^2}{\left(x^2 - y^2\right)^2} - \frac{y^2}{x^2 - y^2} d\Omega^2, \tag{3.6}$$

where

$$d\Omega^2 = d\theta^2 + \sin^2\theta d\varphi^2$$

is the metric on the unit sphere, and k is an arbitrary constant.

Notably, if transformation (3.5) occurs with preservation of the determinant of matrix $M\,(L = R)$:

$$det\ M = x^2 - y^2 = \text{const.},$$

the metric (3.6) becomes

$$-\left(x^2 - y^2\right)ds^2 = dx^2 - dy^2 - y^2 d\Omega^2, \tag{3.7}$$

and looks very similar to the metric of Minkowski space if we take $x = ct, y = r$, where t is the time and r is the radial coordinate.

However, even ignoring such a case, it can be shown that x and y obey transformations that are equivalent to Lorentz transformations. The metric (3.6) is dependent only on the ratio x/y. Indeed, by means of transformation

$$x = X\cosh T,\ y = X\sinh T, \tag{3.8}$$

Equation (3.6) can be written as

$$\frac{ds^2}{k^2} = dT^2 + \sinh^2 T d\Omega^2, \tag{3.9}$$

which proves this statement. Therefore, no matter how the coordinate X transforms, the metric is totally indifferent.

This metric has special importance: it is the "public space" metric or, formally speaking, the spatial part of N. Ionescu Pallas's [4] stationary hyperbolic metric. This fact is expected since the metric (3.7) is conformally Minkowski, whose conformity factor is obtained as a benefit only because it is an absolute metric.

Therefore, using (3.9) as a spatial metric relative to the Ernst equation, we solve this equation for the case of a real potential, independent of the angle. In this case, Ernst's equation [7] reduces to

$$\frac{1}{\sinh^2 T}\left(\sinh^2 T f'\right)' = (f)^2,$$

where the accent denotes the derivative with respect to T, with the following solution

$$f(T) = f_0 e^{-\alpha \cosh T}$$

leading to the hyperbolic stationary metrics:

$$ds^2 = f_0 e^{-a\cosh T} dt^2 - \frac{1}{f_0} e^{a\cosh T}\left(dT^2 + \sinh^2 T d\Omega^2\right), \quad (3.10)$$

t is the temporal coordinate. Matrix (3.10) represents a generalization of that given by N. Ionescu Pallas [4], obtained from (3.10) with $f_0 = 1, a = 0$.

3.4 Directions in Space at Small Scale

We are now interested in coordinate transformations (X, T); since such transformations are connected to transformations in x and y., it is important to set the problem of their finding. To this end, we employ a very much-used viewpoint in modern theoretical physics, namely, considering vectors as abstract quantities satisfying quaternion algebra [8]. Indeed, (3.3) can be used to generally express M in terms of Pauli matrices, while the linear transformations in x and y (only) induce transformations of these matrices, which form an absolute orthogonal frame. The basis of such a frame is usually given by three 2×2 mutually anticommutative matrices, say I, J, K, which satisfy the multiplication table:

$$\begin{gathered} I^2 = J^2 = K^2 = E, I \cdot J \cdot K = iE \\ IJ = -JI = iK, JK = -KJ = iI, KI = -IK = iJ, \end{gathered} \quad (3.11)$$

where $i = \sqrt{-1}$. The "vectors" of this space are 2×2 matrices with null traces (involutions), whose norm is given by the scalar product:

$$(A \cdot B) = \frac{1}{2}(A \cdot B + B \cdot A)$$

which is a multiple of the unit matrix:

$$(MM) = \left(M_1^2 + M_2^2 + M_3^2\right)E. \quad (3.12)$$

Among frames (3.11), the one given by Pauli's matrices is

$$I_0 = \begin{bmatrix} 0 & i \\ -i & 0 \end{bmatrix}, J_0 = \begin{bmatrix} 0 & 1 \\ 1 & 0 \end{bmatrix}, K_0 = \begin{bmatrix} 1 & 0 \\ 0 & -1 \end{bmatrix} \quad (3.13)$$

plays an essential role in the algebraic theory of spin $1/2$.

In such frames, the 2×2 matrices are represented by linear combinations of the unit matrix and the elements (3.11) of the frame. The vector part of the matrix is then given as a linear combination of the last ones only. A null vector does not also mean the null matrix. However, by all means, one or both of its eigenvalues must be zero; according to the canons of quantum mechanics, the physical significance of measurements is based on these elements. These eigenvalues can also be given as, for example, some functions of x and y, which are obviously zero. Therefore, in such an extension of the theory, the eigenvalues are not necessarily given by simple linear expressions such as (3.4).

If we accept the null vectors as being the "cosmic background", that is, the "zero" of the devices, it follows their major importance in vacuum problems, such as of the matrices previously discussed. The geometry of the null vectors was studied in detail by Sobczyk [9], and we used two of his important theorems.

Let us define in frame (3.13) a special null vector associated with the direction of I_0, whose canonical form is:

$$N_0 = (E + J_0) \cdot K_0 = K_0 + iI_0. \tag{3.14}$$

We are interested in those frame transformations that maintain this canonical form, that is, those transformations $(I_0, J_0, K_0) \to (I, J, K)$ that maintain the form of the special null vector associated with the direction of I:

$$N = (E + J) \cdot K = K + iI. \tag{3.15}$$

Sobczyk [9] showed that, in this case, there must be the following transformation:

$$N = e^{\frac{u}{2}N} N_0 e^{-\frac{u}{2}N}, \tag{3.16}$$

where u is a parameter. Writing out this transformation explicitly by means of (3.14) and (3.15), one can easily find the frame transformations that maintain the special form of the null vector:

$$\begin{aligned} I &= \left(1 + \frac{u^2}{2}\right) I_0 + iuJ_0 - i\frac{u^2}{2} K_0, \\ J &= -iuI_0 + J_0 - uK_0, \\ K &= -i\frac{u^2}{2} I_0 + uJ_0 + \left(1 - \frac{u^2}{2}\right) K_0. \end{aligned} \tag{3.17}$$

Another Sobczyk's theorem shows that any null vector of a frame can be written in terms of the special null vector (3.15) in the form:

$$N' = BNB, \tag{3.18}$$

where B is a vector orthogonal to the direction associated with N:

$$B = UJ + VK. \tag{3.19}$$

Form (3.18) is the so-called Cartan–Whittaker representation of the null vector. According to (3.19), it can be written as

$$N' = x_1 I + x_2 J + x_3 K, \tag{3.20}$$

where the components are

$$\begin{aligned} x_1 &= -i\left(U^2 + V^2\right), \\ x_2 &= 2UV, \\ x_3 &= V^2 - U^2, \end{aligned} \tag{3.21}$$

which are defined only up to an arbitrary factor and to a permutation that takes into account the direction of the frame associated with the special null vector.

In relation to transformation (3.17), the quantities (3.21) compose the one-parameter linear group:

$$\begin{aligned} x_1\prime &= \left(1 + \frac{u^2}{2}\right)x_1 - iux_2 - i\frac{u^2}{2}x_3, \\ x_2\prime &= iux_1 + x_2 + ux_3, \\ x_3\prime &= -i\frac{u^2}{2}x_1 - ux_2 + \left(1 - \frac{u^2}{2}\right)x_3, \end{aligned} \tag{3.22}$$

which is equivalent to transformations:

$$\begin{aligned} V' &= V, \\ U' &= U + u \end{aligned} \tag{3.23}$$

of the parameters U and V.

If X and T in formula (3.8) are the components of a spinor (3.19): $X = V, T = U$, then transformation (3.23), in parameters x and y, is expressed by

$$\begin{aligned} x' &= x\cosh u + y\sinh u, \\ y' &= x\sinh u + y\cosh u. \end{aligned} \tag{3.24}$$

Transformation (3.23) was found by Rindler [10] while studying the analogy between Kruskal coordinates and uniformly accelerated motion in Minkowski's universe. Hence, (3.24) is equally connected with Minkowski's coordinates, on the one hand, and with Rindler's and Kruskal's coordinates, on the other, and represents one-parameter transformations of the abstract frames that make the form (3.15) of special null vectors an invariant. Here, we observe that the Kruskal coordinates can be conceived as components of a null vector.

This reality generates some promising analogies, but unfortunately, we cannot go further with the coordinates X and T because T is nondimensional, and furthermore, the null vectors (3.21) constructed with them would be summations of expressions that are dimensionally inhomogeneous. The truth is that one of the coordinates can be adjusted by an adequate factor to be dimensionally correct (as it usually does [4]), but this fact implies additional arbitrariness. In this case, transformation (3.24) would be defined up to an arbitrary factor, being in fact a 2-parameter transformation. Our problem is, therefore, to directly obtain a generalization for more parameters of group (3.24), considering x and y as the U and V components of the spinors (3.19), in which case the components of the null vectors (3.21) become dimensionally homogeneous.

At the same time, we also look for a generalization of the group (3.22). To obtain it, one observes that transformations (3.22) leave unchanged the quadratic form:

$$x_1^2 + x_2^2 + x_3^2 = 0, \tag{3.25}$$

which gives the norm of the null vectors. Considering the formal identity between (3.24) and the Euclidean norm, it is natural to suppose that the most general invariance group is the group with three rotation parameters, whose infinitesimal generators are

$$\begin{aligned} M_1 &= -i\left(x_2\frac{\partial}{\partial x_3} - x_3\frac{\partial}{\partial x_2}\right), \\ M_2 &= -i\left(x_3\frac{\partial}{\partial x_1} - x_1\frac{\partial}{\partial x_3}\right), \\ M_3 &= -i\left(x_1\frac{\partial}{\partial x_2} - x_2\frac{\partial}{\partial x_1}\right), \end{aligned} \tag{3.26}$$

which truly happens. This fact was supported by Yamamoto [11], who observed that the form (3.25) of infinitesimal generators is equivalent to the form resulting from it for

$$x_1 = \rho sin\,\omega, x_2 = -\rho cos\,\omega, x_3 = -i\rho, \tag{3.27}$$

that is,

$$\begin{aligned} M_1 &= \cos\omega\rho\frac{\partial}{\partial\rho} - \sin\omega\frac{\partial}{\partial\omega} \\ M_2 &= \sin\omega\rho\frac{\partial}{\partial\rho} + \cos\omega\frac{\partial}{\partial\omega}, \\ M_3 &= -i\frac{\partial}{\partial\omega}. \end{aligned} \tag{3.28}$$

The action of operators (3.27) on spin eigenfunctions

$$v_+ = \sqrt{\rho}e^{\frac{i\omega}{2}}, v_- = \sqrt{\rho}e^{\frac{-i\omega}{2}} \tag{3.29}$$

reproduces the action of the Pauli matrices according to the following relations:

$$\begin{aligned}\begin{bmatrix} M_1 v_+ \\ M_1 v_- \end{bmatrix} &= \frac{1}{2}\begin{bmatrix} 0 & 1 \\ 1 & 0 \end{bmatrix}\begin{bmatrix} v_+ \\ v_- \end{bmatrix},\\ \begin{bmatrix} M_2 v_+ \\ M_2 v_- \end{bmatrix} &= \frac{1}{2}\begin{bmatrix} 0 & i \\ -i & 0 \end{bmatrix}\begin{bmatrix} v_+ \\ v_- \end{bmatrix},\\ \begin{bmatrix} M_3 v_+ \\ M_3 v_- \end{bmatrix} &= \frac{1}{2}\begin{bmatrix} -1 & 0 \\ 0 & 1 \end{bmatrix}\begin{bmatrix} v_+ \\ v_- \end{bmatrix}.\end{aligned} \tag{3.30}$$

It can be shown in a straightforward manner that the operators (3.28) satisfy the same algebras as Pauli's matrices.

The infinitesimal transformations (3.28) and (3.26) do not tell us very much about this group. To this end, we will look for some finite transformations generated by infinitesimal transformations. In this respect, we set operators (3.28) in a form capable of demonstrating its isomorphism to the Barbilian group [6, 12–16], as considered in the previous chapter. The new operators are given by the following linear combinations:

$$\begin{aligned} X_1 &= M_1 - iM_2,\\ X_2 &= M_3,\\ X_3 &= -(M_1 + iM_2), \end{aligned} \tag{3.31}$$

and satisfy the same structural relations as the infinitesimal generators of the Barbilian group.

Taking α and $\overleftarrow{\alpha}$ as group variables, with $\alpha = v_+$ from (3.29), the operators (3.31) also write

$$Y_1 = \overleftarrow{\alpha}\frac{\partial}{\partial \overleftarrow{\alpha}},\ Y_2 = \frac{1}{2}\left(\alpha\frac{\partial}{\partial\alpha} - \overleftarrow{\alpha}\frac{\partial}{\partial \overleftarrow{\alpha}}\right),\ Y_3 = -\alpha\frac{\partial}{\partial\alpha}. \tag{3.32}$$

To determine the finite transformations generated by these infinitesimal transformations, we proceed to determine the invariant functions of the operator [17]:

$$U = \mu Y_1 + \nu Y_2 + \lambda Y_3,\ \mu,\ \nu,\ \lambda - \text{ constants}$$

which are solutions of the equation $U\psi = 0$:

$$\left(\mu\overleftarrow{\alpha} + \frac{\nu}{2}\alpha\right)\frac{\partial\psi}{\partial\alpha} - \left(\frac{\nu}{2}\overleftarrow{\alpha} + \lambda\alpha\right)\frac{\partial\psi}{\partial\overleftarrow{\alpha}} = 0 \tag{3.33}$$

with the characteristic system:

$$\frac{d\alpha}{\mu\overleftarrow{\alpha} + \frac{\nu}{2}\alpha} = -\frac{d\overleftarrow{\alpha}}{\lambda\alpha + \frac{\nu}{2}\alpha}. \tag{3.34}$$

We now look for another two constants, m and n, so that the linear combination results from (3.34)

$$\frac{m d\alpha - n d\overleftarrow{\alpha}}{m\left(\frac{\nu}{2}\alpha + \mu\overleftarrow{\alpha}\right) - n\left(\lambda\alpha + \frac{\nu}{2}\overleftarrow{\alpha}\right)} = d\tau \tag{3.35}$$

is the total differential. This can happen only under the following conditions:

$$\begin{aligned} &\left(\tfrac{\nu}{2} - \rho\right)m - \lambda n = 0, \\ &\mu m - \left(\tfrac{\nu}{2} - \rho\right)n = 0, \end{aligned} \tag{3.36}$$

where ρ is an arbitrary factor. This system is compatible only with the following values of ρ:

$$\rho_1 = \frac{\nu}{2} + \sqrt{\lambda\mu},\ \rho_2 = \frac{\nu}{2} - \sqrt{\lambda\mu}.$$

For the first value, (3.35) yields the integral:

$$\sqrt{\lambda}\alpha + \sqrt{\mu}\,\overleftarrow{\alpha} = \left(\sqrt{\lambda}\alpha_0 + \sqrt{\mu}\,\underset{0}{\overleftarrow{\alpha}}\right)e^{\left(\frac{\nu}{2} + \sqrt{\lambda\mu}\right)\tau}, \tag{3.37}$$

and for the second, the integral:

$$\sqrt{\lambda}\alpha - \sqrt{\mu}\,\overleftarrow{\alpha} = \left(\sqrt{\lambda}\alpha_0 - \sqrt{\mu}\,\underset{0}{\overleftarrow{\alpha}}\right)e^{\left(\frac{\nu}{2} - \sqrt{\lambda\mu}\right)\tau}, \tag{3.38}$$

where α_0 and $\underset{0}{\overleftarrow{\alpha}}$ are the values of α and $\overleftarrow{\alpha}$ for $\tau = 0$, respectively. Relations (3.36) and (3.37) give

$$\begin{aligned} \alpha &= e^{\frac{\nu}{2}}\left(\alpha_0 \cosh\sqrt{\lambda\mu}\tau + \sqrt{\tfrac{\mu}{\lambda}}\,\underset{0}{\overleftarrow{\alpha}}\sinh\sqrt{\lambda\mu}\tau\right), \\ \overleftarrow{\alpha} &= e^{\frac{\nu}{2}}\left(\alpha_0\sqrt{\tfrac{\lambda}{\mu}}\sinh\sqrt{\lambda\mu}\tau + \underset{0}{\overleftarrow{\alpha}}\cosh\sqrt{\lambda\mu}\tau\right). \end{aligned} \tag{3.39}$$

For (3.39) to make sense, it is necessary that λ be the complex conjugate of μ : $\overleftarrow{\lambda} = \mu = \gamma e^{i\phi}$, while ν has to be real—and we take it as zero. Therefore, the finite transformations of group (3.31), generated by infinitesimal transformations, are given by the unimodular group:

$$\begin{aligned} \alpha &= \alpha_0 \cosh\gamma\tau + \underset{0}{\overleftarrow{\alpha}}\,e^{i\phi}\sinh\gamma\tau, \\ \overleftarrow{\alpha} &= \alpha_0 e^{-i\phi}\sinh\gamma\tau + \underset{0}{\overleftarrow{\alpha}}\cosh\gamma\tau. \end{aligned} \tag{3.40}$$

It is no accident that we have denoted by α, the group variable. This notation is commonly used to denote the complex amplitude of harmonic oscillators, which is the eigenvalue of the annihilation operator in the second quantization [18]. Transformation (3.40) was first described by Stoler [19] in connection with the generalization of the eigenstates of the annihilation operator (coherent states). For example, if α_0 describes an oscillator in a state satisfying the minimum conditions of momentum coordinate uncertainty, the state α is characterized by the following uncertainty relation:

$$(\Delta p)^2(\Delta q)^2 = \frac{1}{4}\left(1 + \sinh^2 \gamma\tau \sin^2 \phi\right), \tag{3.41}$$

where (Δp) and (Δq) are variations in the momentum and coordinate, respectively, of the oscillator, while the Planck constant is taken to be equal to unity.

Consequently, the variable α can be considered the complex amplitude of a bosonic field. The fact that group (3.31) is isomorphic with the Barbilian group shows that the latter is equally entitled to refer to the variables connected with such a field, and this fact—as we shall prove—actually happens.

The group induced by (3.40) on variables x and y can be obtained from (3.27) by means of the following identifications:

$$\begin{aligned} x_1 &= 2xy, \\ x_2 &= y^2 - x^2, \\ x_3 &= -i\left(x^2 + y^2\right) \end{aligned} \tag{3.42}$$

leading to

$$\begin{aligned} x &= \sqrt{\rho}\cos\tfrac{\omega}{2}, \\ y &= \sqrt{\rho}\sin\tfrac{\omega}{2}. \end{aligned} \tag{3.43}$$

As one can see, $\alpha_0 = x + iy$, and introducing $\alpha' = x' + iy'$ into (3.40), one obtains

$$\begin{aligned} x' &= (\cosh\gamma\tau + \sinh\gamma\tau\cos\phi)x + \sinh\gamma\tau\sin\phi y, \\ y' &= \sinh\gamma\tau\sin\phi x + (\cosh\gamma\tau - \sinh\gamma\tau\cos\phi)y. \end{aligned} \tag{3.44}$$

These transformations reduce to transformations (3.24) for $\phi = \frac{\pi}{2}$ and $\gamma\tau = u$, but the transformations induced on the Cartan-Whittaker component (3.42) do not reduce to those given by relations (3.22). We are particularly interested in the physical significance of spinorial variables $\left(\alpha, \overleftarrow{\alpha}\right)$: each of them signifies a bosonic field. For details concerning the second quantization, Stoler's transformation, and generalized coherent states, the reader is invited to consult [19]. Currently, we realize that treatment of the angular variables on a level with measured coordinates directly leads to spatial significance through the stationary hyperbolic metric of N. Ionescu Pallas;

while making abstraction of angular variables and considering only an "internal" group, the measured quantities signify some field amplitudes.

3.5 The "Motion" Geometry

The original theory of wave-corpuscle duality proposed by de Broglie is based on a theorem that is implicitly contained in the Lorentz transformation [20]. This theorem connects the cyclic frequency of the local horologes at each point of a spatial domain with the progressive wave frequency in phase with the horologes. This wave shows the distribution of the oscillator phases in the considered spatial domain [20]. Let us now show that such a distribution—in the true sense of the word—could be given without appealing to the LT.

De Broglie's idea that all horologes have the same cyclic frequency could be materialized through a periodical field, which, as is well known, is describable by means of local oscillators. The result discussed in the previous paragraph regarding the isomorphism between the Barbilian group and Stoler's centro-affine unimodular group shows, as we have mentioned, that the Barbilian group could be successfully applied to complex field amplitudes.

To prove this idea, we use the harmonic oscillator equation:

$$x'' + \omega^2 x = 0, \tag{3.45}$$

where x is the field-relevant coordinate and ω is the cyclic frequency. This equation represents a model for any harmonic of the field. The most general solution of (3.45) depends on three constants and can be written as

$$x(t) = he^{i(\omega t+\varphi)} + h^* e^{-i(\omega t+\varphi)}, \tag{3.46}$$

where the complex amplitude h is most important in the coherent theory of bosonic fields [18]. The quantities h and h^* display the initial conditions, which differ from one spatial point to another. In other words, at a given moment in time, the oscillators situated at various points in space are in different quantum states and have different phases. There is a problem that logically appears: is it possible to provide an a priori connection between the parameters h and h^*, on the one hand, and $e^{i(\omega t+\varphi)}$, on the other hand, characterizing various oscillators at a certain moment of time? Since (3.46) is a solution of (3.35), the answer is affirmative: formula (3.45) possesses a "hidden" symmetry, expressed by the homographic group: the ratio $\tau(t)$ of two solutions of Eq. (3.45) is a solution of Schwartz's equation [1]:

$$\left(\frac{\tau''}{\tau'}\right) - \frac{1}{2}\left(\frac{\tau''}{\tau'}\right)^2 = 2\omega^2. \tag{3.47}$$

This equation is invariant with respect to a homographic transformation of $\tau(t)$: any homographic function of τ is, in turn, a solution of Eq. (3.47). Since homography is characteristic of projectivity on a straight line, we can affirm that the ratio of two solutions of Eq. (3.45) is a projective parameter for the multitude of oscillators of the same frequency in a given spatial domain.

Now, one can easily construct a convenient projective parameter that is in biunivocal correspondence with the oscillator. In this respect, we first observe a "universal projective parameter", which is the ratio of the fundamental solutions of (3.45):

$$k = e^{2i(\omega t + \varphi)}. \tag{3.48}$$

Any homographic function of this ratio represents a new projective parameter. One of these functions, namely:

$$\tau(t) = \frac{h + h^* k}{1 + k} \tag{3.49}$$

has the advantage of being specific to each oscillator. In addition, let us consider another function

$$\tau' = \frac{h' + h^{*'} k'}{1 + k'}, \tag{3.50}$$

which is specific to another oscillator. Since (3.49) and (3.50) are solutions of Eq. (3.47), the homographic relation exists between them:

$$\tau' = \frac{a\tau + b}{c\tau + d}.$$

Written explicitly, these relations lead to group equations:

$$\begin{aligned} h' &= \tfrac{ah+b}{ch+d}, \\ k' &= \tfrac{ch^*+d}{ch+d} k. \end{aligned} \tag{3.51}$$

The structure of this group is typical of an SL(2, R) group, which we take in the standard form:

$$[B_1, B_2] = B_1, [B_2, B_3] = B_3, [B_3, B_1] = -2B_2, \tag{3.52}$$

where B_k are the infinitesimal generators of the group. Because the group is simply transitive, these generators can be easily found as the components of the Cartan frame [21–23] from the formula

$$d(f) \equiv \sum \frac{\partial f}{\partial x^k} dx^k$$
$$= \left[\omega^1\left(h^2\frac{\partial}{\partial h} + h^{*2}\frac{\partial}{\partial h^*} + (h - h^*)k\frac{\partial}{\partial k}\right) + 2\omega^2\left(h\frac{\partial}{\partial h} + h^*\frac{\partial}{\partial \bar{h}}\right) + \omega^3\left(\frac{\partial}{\partial h} + \frac{\partial}{\partial h^*}\right)\right] \tag{3.53}$$

where ω^k are the components of the Cartan coframe to be found from the system:

$$dh = \omega^1 h^2 + 2\omega^2 h + \omega^3,$$
$$dh^* = \omega^1 h^{*2} + 2\omega^2 h^* + \omega^3$$
$$dk = \omega^1 k(h - h^*).$$

Thus, we immediately obtain both the infinitesimal generators and the coframe by identifying the right-hand side of Eq. (3.53) with the standard dot product of se (2, R) algebra:

$$\omega^1 B_3 + \omega^3 B_1 - 2\omega^2 B_2$$

so that

$$B_1 = \tfrac{\partial}{\partial h} + \tfrac{\partial}{\partial h^*},\ B_2 = h\tfrac{\partial}{\partial h} + h^*\tfrac{\partial}{\partial h^*}$$
$$B_3 = h^2\tfrac{\partial}{\partial h} + h^{*2}\tfrac{\partial}{\partial h^*} + (h - h^*)k\tfrac{\partial}{\partial k}$$

and

$$\omega^1 = \frac{dk}{(h - h^*)k},\ 2\omega^2 = \frac{dh - dh^*}{h - h^*} - \frac{h + h^*}{h - h^*}\frac{dk}{k}$$
$$\omega^3 = \frac{hdh^* - h^*dh}{h - h^*} + \frac{hh^*}{h - h^*}\frac{dk}{k} \tag{3.54}$$

In real terms, $h = u + iv, k = e^{i\phi}$, and these last equations can be written as

$$B_1 = \tfrac{\partial}{\partial u},\ B_2 = u\tfrac{\partial}{\partial u} + v\tfrac{\partial}{\partial v},\ B_3 = \left(u^2 - v^2\right)\tfrac{\partial}{\partial u} + 2uv\tfrac{\partial}{\partial v} + 2v\tfrac{\partial}{\partial \phi}$$
$$\omega^1 = \tfrac{d\phi}{2v}\ \ \omega^2 = \tfrac{dv}{v} - \tfrac{u}{v}d\phi\ \ \omega^3 = \tfrac{u^2+v^2}{2v}d\phi + \tfrac{vdu-udv}{v}. \tag{3.55}$$

It should be mentioned that, in his original paper, Barbilian does not work with the above differential forms but with the absolute invariant differentials

$$\omega^1 = \frac{dh}{(h - h^*)k},\ \omega^2 = -i\left(\frac{dk}{k} - \frac{dh + dh^*}{h - h^*}\right),\ \omega^3 = -\frac{kdh^*}{h - h^*} \tag{3.56}$$

or, in real terms, exhibiting a three-dimensional Lorentz structure of this space

$$\Omega^1 \equiv \omega^2 = d\phi + \frac{du}{v},$$
$$\Omega^2 = \cos\phi\frac{du}{v} + \sin\phi\frac{dv}{v},\ \Omega^3 = -\sin\phi\frac{du}{v} + \cos\phi\frac{dv}{v}. \tag{3.57}$$

The advantage of this representation is that it makes obvious the connection with the Poincaré representation of the Lobachevsky plane. Indeed, the metric here is

$$-\left(\Omega^1\right)^2 + \left(\Omega^2\right)^2 + \left(\Omega^3\right)^2 = -\left(d\phi + \frac{du}{v}\right)^2 + \frac{(du)^2 + (dv)^2}{v^2}$$
$$= \left(\frac{dk}{k} - \frac{dh + dh^*}{h - h^*}\right)^2 - 4\frac{dhdh^*}{(h - h^*)^2}. \tag{3.58}$$

This metric reduces to that of Poincaré in the case where $\Omega^1 = 0$, which, as Barbilian noted, defines the variable ϕ as the "angle of parallelism" of the hyperbolic plane (the connection). In fact, in modern terms, (du/v) represents the connection form of the hyperbolic plane [23], and Eq. (3.57) represents a general Băcklund transformation in that plane [24, 25].

In view of the importance that we revealed of the geometry of Lobachevsky in the classical Kepler problem, it becomes important to know the meaning of the condition $\Omega^1 = 0$ for a family of cubic equations. It turns out that it expresses the so-called apolar transport of cubics [26], whereby a certain cube evolves in such a way that its roots always remain in a harmonic progression. This transport is defined by the condition that any root of the "transported" cubic is in a harmonic relation with any root of the "original" cubic with respect to the other two remaining roots of the original cubic

$$\frac{y_l - x_j}{y_l - x_k} : \frac{x_i - x_j}{x_i - x_k} = -1; i \neq j \neq k \neq i \tag{3.59}$$

in all positive permutations of the indices i, j, k and for every l. Therefore, each new root $\left(y_l\right)$ and each of the corresponding old ones (x_i) are in the harmonic range with respect to the other two old roots $\left(x_j, x_k\right)$. Then, it can be proven that the conditions from Eq. (3.59) boil down to the vanishing of the bilinear invariant of the two cubics, analogous to the bilinear invariant of the quadratics:

$$a_3b_0 - 3a_2b_1 + 3a_1b_2 - a_0b_3. \tag{3.60}$$

Here, a_m denotes the coefficient of the original cubic layer, while b_m denotes the coefficient of the transported cubic layer. Obviously, this invariant is analogous to the one from the case of two quadratics, whose vanishing expresses that their roots are in a harmonic sequence. The geometry related to this invariant is century old [27], and Dan Barbilian seemed particularly fond of it [26] because he elaborated for a long time on its different aspects, especially related to the geometry of the triangle. As triangles currently exist in relation to the construction of skyrmions from instantons [28, 29], from a point of view closely related to their geometry, it is, therefore, worth considering this connection, which is strictly related to the physics of continua.

Now, if the two cubes are infinitesimally close, then the condition of their transport by involution reduces to

$$a_3 da_0 - 3a_2 da_1 + 3a_1 da_2 - a_0 da_3 = 0. \tag{3.61}$$

Here, the correspondences between the coefficients of the cubic

$$a_0 x^3 + 3a_1 x^2 + 3a_2 x + a_3 = 0$$

and the quantities h,h^*, k, ie

$$a_0 = 1 + k^3, a_1 = -\left(h + h^* k^3\right), a_2 = h^2 + h^{*^2} k^3$$

above for the coefficients, the condition of apolar transport of the cubic amounts to

$$-\left(h - h^*\right)^3 \cdot k^3 \left(\frac{dk}{k} - \frac{dh + dh^*}{h - h^*}\right) = 0.$$

As the cubes are assumed to have distinct roots, this condition is satisfied if, and only if, the differential form Ω^1 is null. Therefore, the parallel transport of the hyperbolic plane actually represents the apolar transport of the cubics.

In this way, the vectors represented by the real eigenvalues of a certain matrix can be physically interpreted in the framework of the classical theory of Kepler motion. For the sake of example, assume that a hydrogen atom is described by a single Kepler motion. The nuclear matter in such a representation can be characterized by a complex number depending on the eccentricity and orientation of the orbit. This complex number can be assumed to represent a particular state of stress inside the nucleus.

The mechanism of generation of the metric of the stationary field is now clearly understood through the methodology of the Matzner–Misner variational principle. If we describe this field by the amplitude and phase of its harmonics at a certain moment in time, then these oscillators form assemblies of a given frequency, which are transitivity varieties of the Barbilian group (de Broglie coherence in Barbilian sense). The invariant metric of this group generates the Lagrangian leading to Ernst's equations [7].

It is important to mention, in this context, that there also exists a transitivity group between assemblies of various frequencies: we meet, this way, the Stoler group, as discussed in the previous paragraph. This fact is easily observed since the parameter $(\gamma\tau)$ of the Stoler transformation is precisely the frequency ratio when the creation and annihilation operators refer to a harmonic oscillator.

We are now capable of giving the equivalent of the de Broglie wave function, that is, the distribution of phases and amplitudes on an assembly of given frequencies. As we observed, the Barbilian group is measurable, with the elementary measure given by

$$dP = -\frac{dh \wedge dh^* \wedge dk}{(h - h^*)^2 k}. \tag{3.62}$$

Therefore, the repartition density of such an elementary probability is the invariant function of the group:

$$\frac{1}{(h - h^*)^2 k}. \tag{3.63}$$

Typically, the square root of this function is defined as an arbitrary unimodular factor and can be considered analogous to the wave function. However, this time, it does not satisfy the Schrödinger equation but rather some equations derived from Ernst's equations (see Chap. 2). Nevertheless, taking into account these equations, one can deduce an important fact: the "wave" function

$$\psi = \frac{1}{v} e^{i\phi/2} \tag{3.64}$$

obtained from (3.53), satisfies the stationary "Schrödinger type equation"

$$\nabla^2 \psi = \left\{ \left(\frac{\nabla v}{v} \right)^2 + \frac{\nabla u \nabla v}{v} - \frac{(\nabla \phi)^2}{4} + \frac{1}{2} i \frac{\nabla \phi \nabla v}{v} \right\} \psi \tag{3.65}$$

with a complex "eigenvalue". This becomes real if either the oscillator assembly has the same (spatial) phase or the same amplitude. In the first case, which is very interesting, Eq. (3.65) reduces to

$$\nabla^2 \psi = \left(\frac{\nabla v}{v} \right)^2 \psi. \tag{3.66}$$

Its "eigenvalue" is the squared "momentum" responsible for the coordinate-momentum uncertainty in the hydrodynamic model of quantum mechanics. In fact, both Ernst's equations and the way we define the wave function show that a hydrodynamic model of such "quantum mechanics" is more easily approached [13, 30].

This investigation offers an example of how the a priori generated probabilities, as measures of some parametric groups, can interfere with fundamental physical theories [31]. Due to the importance of Ernst's equation, the Barbilian group seems to be essential in explaining the connection between Relativity and Undulatory Mechanics.

On the other hand, the metric (3.58) now offers a natural possibility of extension of the harmonic principle that we associated with a Kepler problem by a functional that, in fact, represents the deformation of the nuclear matter in the general case. The extension amounts to expressing the energy of mapping by the functional

$$E(\boldsymbol{\Phi}) = \iiint \left\{ -4\frac{(\nabla h \cdot \nabla h^{*})}{(h - h^{*})^{2}} + \left(\frac{dk}{k} - \frac{\nabla h + \nabla h^{*}}{h - h^{*}} \right)^{2} \right\} (d^{3}\boldsymbol{x}). \tag{3.67}$$

The variational principle would then express this simple fact: the most general deformation of the nuclear matter is replicated in the atomic structure by the variation of the eccentricity of the electronic orbits. Likewise, inside the nucleus, the variation is replicated by an ensemble of harmonic oscillators of the same frequency. The second term from this functional is analogous to the baryonic term from the theory of Skyrme, with the only difference being that the whole functional is homogeneous. In the language of skyrmion technology, however [32], it still represents hyperbolic skyrmions because the geometrical character of the problem is dictated by the metric from Eq. (3.58).

In this form, the theory illustrates the strong ties of the model of nuclear matter with the ideas of confinement of matter in general. To show this, we will present a solution to the variational principle related to the energy functional from Eq. (3.67). To better understand this solution, let us start with "strange" classical dynamics, which are related to the forces involved in the problem of confinement of the classical ideal gas and are inversely proportional in magnitude to the distance between molecules. This force is also involved in a Newtonian description of inertia [33], which, again, should appear to be quite natural in view of the common background of general relativity and Skyrme theory.

Assuming a Newtonian dynamic for this force, the equations of motion can be written in the following form:

$$\overset{\prime}{\vec{r}} + \frac{\mu^{2}}{r^{2}}\vec{r} = \vec{0}.$$

This is a plane motion because the force is central. In polar coordinates of the plane of motion, the equations of motion are split into a system of two differential equations

$$r' - r\theta'^{2} + \frac{\mu}{r} = 0, \; r'\theta' + 2r'\theta' = 0 \tag{3.68}$$

The second of these gives the area constant

$$r^{2}\theta' = a'.$$

The first Eq. (3.68) can then be integrated as follows [34]: first, we change the time variable into

$$\tau(t) = \int \frac{dt}{r^{2}} \therefore \theta'(\tau) = \overset{\prime}{a}, \tag{3.69}$$

where the prime represents differentiation with respect to τ. Now, if we change the dependent variable to $\xi(\tau) = 1/\mathrm{r}$, the first Eq. (3.68) becomes

$$\xi'' + \left(\overset{\prime 2}{a} - \mu^2\right)\xi = 0; \xi(\tau) \equiv \frac{1}{r} \tag{3.70}$$

This is the equation of a harmonic oscillator, with the frequency dictated by the rate of area, leading to either trigonometric or hyperbolic functions. We consider only the first case, when the solutions are of the form

$$\xi(\tau) = A\sin(\Omega\tau) + B\cos(\Omega\tau),\ \Omega^2 \equiv \overset{\prime 2}{a} - \mu^2 > 0. \tag{3.71}$$

Now, we can find the relation between the Newtonian time and the new time τ. Taking Eq. (3.71) into Eq. (3.69) gives

$$t = \int \frac{d\tau}{\xi^2(\tau)} = \int \frac{d\tau}{[A\sin(\Omega\tau) + B\cos(\Omega\tau)]^2} \tag{3.72}$$

and this integral leads to [35]

$$\Omega\tau = \phi_0 + \tan^{-1}\left[\left(A^2 + B^2\right)\Omega t\right], \tag{3.73}$$

where ϕ_0 is a constant of integration. Interestingly, Eq. (3.71) represents a particular Kepler motion, corresponding to a null gravitational constant, null mass, or even null charge. It cannot therefore be interpreted in terms of the motion of a material point around another attracting material point. However, it can be interpreted in terms of abstract kinematics.

The kinematics generated by differential forms (3.57) are considered. In terms of these, the metric (3.58) assumes the Lorentzian form (3.57). Along the geodesics of this metric, the rates represented by the differential forms (3.57) are constant, so we can find those geodesics from some differential equations involving a linear parameter in the arclength from Eq. (3.58). These are

$$\begin{aligned} &d\phi + \frac{du}{v} = a \cdot dt, \\ &\cos\phi\frac{du}{v} + \sin\phi\frac{dv}{v} = b \cdot dt, -\sin\phi\frac{du}{v} + \cos\phi\frac{dv}{v} = c \cdot dt, \end{aligned} \tag{3.74}$$

where a, b, and c are constants. The last two of these equations give

$$\frac{u'}{v} = b\cos\phi - c\sin\phi,\ \frac{v'}{v} = b\sin\phi + c\cos\phi \tag{3.75}$$

and then from the first of (3.74), we have

$$\overset{\prime}{\phi} = \text{a} - \text{bcos}\phi + \text{csin}\phi \tag{3.76}$$

an equation that can be integrated immediately. We prefer to perform this integration by putting the right-hand side of (3.76) in the form of a perfect square to show that this corresponds to Eq. (3.72) above. Indeed, if we take $2\Omega\tau \equiv \phi$, we can write the integral in the form (3.73) for a, b, c given by

$$a \equiv \frac{A^2 + B^2}{2}, b \equiv \frac{A^2 - B^2}{2}, c \equiv AB. \tag{3.77}$$

This gives an interpretation to the classical "time" variable τ, provided we know something about this abstract kinematics. With respect to the parameters u and v, using Eq. (3.75), we have

$$\frac{v}{v_0} = a - b\cos\phi + c\sin\phi, \tag{3.78}$$

where v_0 is another integration constant. Therefore, v represents the inverse square of the position vector of the motion previously described. On the other hand, for the parameter u, we find the following solution

$$v_0(u - u_0) = b\sin\phi + c\cos\phi, \tag{3.79}$$

where u_0 is another integration constant.

Now, one can find a particular solution of the variational principle applied to the energy functional (3.67) along the geodesics given by Eqs. (3.78) and (3.79) if we assume that their parameter (and therefore the phase ϕ) is a solution of the Laplace equation. This can be proven, the easiest way, by continuing to work on real parameters u, v, and ϕ, as before. The Euler–Lagrange equations associated with the variational principle applied to (3.67) are

$$\nabla \cdot \left(\nabla\phi + \frac{\nabla \text{u}}{\text{v}} \right) = 0; \nabla^2\phi - \nabla\phi \cdot \frac{\nabla \text{v}}{\text{v}} = 0; \nabla^2(\ln \text{v}) - \nabla\phi \cdot \frac{\nabla \text{u}}{\text{v}} = 0. \tag{3.80}$$

On the other hand, the geodesics of the metric (3.58) are solutions of the system of differential equations

$$\left(\frac{\text{u}\prime}{\text{v}}\right)\prime + \phi' \cdot \frac{\text{v}\prime}{\text{v}} = 0; \phi'' - \phi' \cdot \frac{\text{v}\prime}{\text{v}} = 0; (\ln \text{v})'' - \phi' \cdot \frac{\text{u}\prime}{\text{v}} = 0, \tag{3.81}$$

where the prime symbol indicates differentiation with respect to the geodesic parameter. Now, if the functions u, v, and ϕ depend on position only through the geodesic parameter, Eq. (3.80) can be written in the form

$$\left(\phi' + \frac{u'}{v}\right)(\nabla t)^2 + \left(\phi' + \frac{u'}{v}\right)\nabla^2 t = 0;\ \left(\phi' - \phi' \cdot \frac{v'}{v}\right)(\nabla t)^2 + \phi \nabla^2 t = 0$$

$$\left((1nv)' - \phi' \cdot \frac{u'}{v}\right)(\nabla t)^2 + (1nv)'\nabla^2 t = 0$$

The first terms of these equations are zero as a consequence of the geodesic equations. Therefore, we still have another form of the harmonic principle: under fairly general conditions, the harmonic mapping corresponding to energy (3.67) is given by the geodesics of the metric, provided that their parameter is a regular harmonic function. This of course makes the phase ϕ the arctangent of such a function, in view of Eq. (3.76) above.

Therefore, at least in this particular instance, we have to focus on the geodesics of the metric (3.58). The Killing vectors of the metric represent conservation laws, and thus, the parameters a, b, and c from Eqs. (3.78) and (3.79) represent these conservation laws. Equations (3.78) and (3.79) represent two Kepler motions, and we are certainly in position to know exactly the field of application of the above kinematics: the theory of space stresses involved in the Kepler problem. These stresses are induced in the core of the solar system, for instance, or in the nucleus, and they represent a material point—in the sense of Hertz—whose particles, acted upon by the inverses of the space elastic forces, behave like an ideal gas. The density of this material point varies inversely proportional to the square of the distance from the origin of the reference frame, which is actually its origin. We must insist upon these two aspects of the problem of action at a distance because they are instrumental in understanding how the nucleus works.

One of the first theories of nuclear matter was that of a gas model [36]. It was not so successful but certainly touched on a fundamental aspect of the problem of the structure of nuclear matter, which actually turns out to be the fundamental problem regarding the structure of matter in general. In view of the above discussion, we may assume the following scenario: the particles of the nucleus are decided "in pairs" by the inversion transformation between external Newtonian forces and the forces, internal to the nucleus, of confinement. These latter ones are represented as harmonic oscillators—the gluons as they were—and described by a general dynamic. The kinematics of a pair are represented as two Kepler motions given by Eqs. (3.78) and (3.79) above. Here, of course, we assume that the pair is "accidentally decided" by inversion, but once decided, it is described by an actual state of stress whose kinematics can be classically described. This state of stress can be physically characterized by a statistic of the kind specifically connected with the light [14].

With Eqs. (3.78) and (3.79) above, we can certainly extend to matter, particularly to the nucleus, Mac Cullagh's view of the structure of light [37]. The two equations represent two harmonic oscillators and two Kepler motions. In classical terms, one can say that two particles (partons) inside the nucleus have independent motions of a Kepler kind. Even though classically described, such an image should be statistically accessible for measurement through some kind of stresses or strains and reflected in the eccentricity of electronic orbits.

This image of the structure of the nucleus can be made even more precise, from the classical point of view we advocate in this work. Indeed, inasmuch as we are thinking in the framework of Newtonian natural philosophy, the above view on Hertz's particles—the partons of nuclear matter—recovers and solves, in a "quantum–mechanical" way, one of the most important issues of that philosophy, left behind by Newton in his theory of forces.

Finally, we cannot ignore the fact that the Barbilian group was first defined as a covariant group of binary cubic forms [12]. In our context, this fact can have a special significance.

Keeping in mind that a horologe has to be attached to any physical frame, each horologe is at the same time an oscillator (periodical motions), so that it is clear that the Barbilian group can be used for synchronization. Let us consider a spatial region endowed with "standard" horologes, which are circular motions with unique frequencies ω. Since the method of considering the frame is arbitrary, we choose all frames by Cartesian axes under the condition that the trajectory of circular motion representing the horologe is in the plane:

$$x + y + z = 0.$$

The oscillator motion is then equivalent to three oscillatory motions on the axes, having equal amplitudes and being de-phased from each other by $\pi/3$. The projective parameters (3.49) of this motion are solutions of a certain cubic equation given by Barbilian [12].

However, it is known that, in general, three solutions of a cubic variable can be considered spatial coordinates [1]: they are the so-called generalized elliptic coordinates. In such coordinates, the variables h, h^*, k are expressed rationally, at most algebraically. This may be the hidden reason that, in these coordinates, the Laplace equation always allows for the separation of variables.

Among the significances of the Barbilian group, we cannot neglect the fact that it was identified for the first time as a covariance group of binary cubic forms. This property, in addition to those associated with propagation and synchronization, is overwhelmingly affected by its implications. Indeed, the physical frame is given as a system of points that are relatively fixed. They also have to be fixed—for example—with respect to the field propagation velocities inside the frame (which is why we emphasize "relatively fixed"). For this reason, the definition of the frame must not contain the displacement mobility. In our opinion, this fact illustrates the profound significance of de Broglie's theorem [20] regarding the equivalence between the rectilinear motion and propagation of a wave: anything implying displacement can be described by a field variable. In this way, we arrive at the qualitative graduation of possible motions of the points of a frame: there are motions that cannot be described by field variables, i.e., they are extremely slow with respect to the temporal scale of processes described in the frame. We can speak, therefore, about frame deformations.

For example, the galaxy, which is taken as the frame for solar system phenomena, is deforming, and considering this deformation, one can explain many negative aspects of the gravitation theory [13, 14]. Moreover, the notion of a frame can also be used in microcosmos: the crystalline net of a metal is a physical frame for the internal processes of the metal (for example, the motion of electrons), but it deforms proprio motu. Even if no force acts on the crystal, the crystalline net is deformed by creep.

The deformation is described by a matrix 3×3, which is the deformation matrix. In irreversible (plastic) deformations, not the matrix elements themselves are important but rather some local invariants constructed with them, that is, the coefficients of the secular equation of the respective matrix:

$$\lambda^3 - I_1\lambda^2 + I_2\lambda - I_3 = 0. \tag{3.82}$$

The roots of this equation are the eigenvalues of the deformation matrix—the main deformations.

Assuming that a frame deformation is described by such a matrix, its invariants are point functions on the frame, as well as the invariants of (3.58). In this way, regarding the inexistence of a fixed frame and its deformation being caused by the gravitational field, we want to describe this deformation as a point function on the frame in a convenient coordinate system and then use it to construct the gravitational field, which produces it. The recipe is offered by the Barbilian group, which is the covariance group of the cubics (3.58). The main deformations can be written in the form given by Barbilian:

$$\varepsilon_i = \frac{h + \omega_i h^* k}{1 + \omega_i k}, \tag{3.83}$$

where $\omega_i^3 = 1$, while h, h^*, and k are algebraic functions of I_1, I_2, I_3. The cubic variation (3.82) at every point, i.e., the evolution of deformation at every point, can now be described by the Barbilian group with variables h, h^*, k. It then follows that by applying the Matzner–Misner principle, we can generalize a stationary metric describing the gravitational field through frame deformation. In our opinion, this is the main reason for reducing the description of the gravitational field to Ernst's equation (see Chap. 2).

Here, we directly observe that two temporal measures exist: there is no need for postulation. The matter concerns, on the one hand, the time of the quadri-dimensional stationary metric and the time of phenomena describable in the frame on the other. These are analogous to Milne's temporal measures. However, we still do not know whether or not some kind of transformation between them exists. The answer to this problem could be essential in the generalization of the method to nonstationary gravitational fields, even if Ernst himself [7] is skeptical about this possibility.

References

1. Mihăileanu, N. (1972). Analytical, differential and projective geometry. Bucharest: Complements, Didactic and Pedagogical Publishing House.
2. Gogala, B. (1980). *International Journal of Theoretical Physics, 19*, 573.
3. Dumitru, S. (1984). *Microphysics (in Romanian)*. Cluj-Napoca: Dacia Publishing House.
4. Ionescu-Pallas, N. (1980). *General relativity and cosmology (in Romanian)*. Bucharest: Scientific and Encyclopedic Publishing House.
5. Schwartz, H. M. (1977). *International Journal of Theoretical Physics, 16*, 249.
6. Barbilian, D. (1968, 1971, 1974), Opera Didactica (in Romanian), Vol. I, II, III. Bucharest: Technical Publishing House.
7. Ernst, F. J. (1971). *Journal of Mathematics and Physics, 12*, 239.
8. Casanova, G. (1979). *L'Algèbre Vectorielle*. Moscou: Mir.
9. Sobczyk, G. (1984). *Acta Physica Polonica B, 12*, 509.
10. Rindler, W. (1966). *American Journal of Physics, 34*, 1174.
11. Yamamoto, T. (1952). *Progress of Theoretical Physics (Japan), 8*, 258.
12. Barbilian, D. (1967). *Opera Mathematica (in Romanian)*, vol. I, p. 286. Bucharest: Romanian Academy Publishing House.
13. Mercheş, I., & Agop, M. (2016). *Differentiability and fractality in dynamics of physical systems*. World Scientific.
14. Mazilu, N., & Agop, M. (2012). *Skyrmions. A great finishing touch to classical Newtonian philosophy*. Nova, New-York: World Philosophy Series.
15. Gottlieb, I., Agop, M., Buzdugan, M., & Crăciun, P. (2005). El Naschie's Cantorian frames, gravitation and quantum mechanics. *Chaos, Solitons and Fractals, 24*, 391–405.
16. Agop, M., Griga, V., Ciobanu, B., Buzea, C., Stan, C., & Tatomir, D. (1997). The uncertainty relation for an Assembly of Planck-Type Oscillations. A possible GR-Quantum Mechanics Connection Chaos. *Solitons and Fractals, 8*(5), 809–821.
17. Vrânceanu, G. (1979). *Lectures on differential geometry (in Romanian)*, Vol. III. Bucharest: Didactic and Pedagogical Publishing House.
18. Carruthers, P., & Nieto, M. M. (1968). *Reviews of Modern Physics, 40*, 411.
19. Stoler, D. (1970). *Physical Review, D, 1*, 3217.
20. De Broglie, L., & La,. (1966). *Thermodinamique de la particule isol´ee*. Paris: Gauthier-Villars.
21. Fels, M., & Olver, P. J. (1998). *Moving Coframes I*, Acta Applicandae Mathematicae, Vol. 51, pp. 161–213.
22. Fels, M., & Olver, P. J. (1999). *Moving Coframes II*, Acta Applicandae Mathematicae, Vol. 55, pp. 127–208.
23. Flanders, H. (1989). *Differential forms with applications to the physical sciences*. New York: Dover Publications.
24. Sasaki, R. (1979). *Soliton equations and pseudospherical surfaces*, Nuclear Physics, Vol. B154, pp. 343–357.
25. Rogers, C., & Schief, W. K. (2002). *Bäcklund and darboux transformations*. Cambridge, UK: Cambridge University Press.
26. Barbilian, D. (1938). *Riemannsche Raum Cubischer Binärformen*, Comptes Rendus de l'Académie Roumaine des Sciences, Vol. 2, p. 345.
27. Burnside, W. S., & Panton, A. W. (1960). *The theory of equations*, Dover Publications.
28. Atiyah, M. F., & Manton, N. S. (1989). Skyrmions from Instantons. *Physics Letters B, 222*, 438–442.
29. Manton, N. S. (1989). Skyrme fields and instantons. In S. K. Durham, & C. B. Donaldson Thomas (Eds.), *Proceedings of the LMS Symposium on Geometry of Low-Dimensional Manifolds* (pp. 179–190), Vol. 1. Preprint DAMTP/89–43.
30. Wilhelm H. E. (1970). *Physical Review, D, 1*, 2278.
31. Jaynes, E. T. (1973). *Foundations of Physics, 3*, 477.
32. Atiyah, M. F., & Sutcliffe, P. (2001). *The geometry of point particles*, arXiv:hep-th/0105.1794v2

33. Sciama, D. W. (1969). *The physical foundations of general relativity*. New York: Doubleday.
34. Govinder, K. S., Athorne, C., & Leach, P. G. L. (1993). The algebraic structure of generalized Ermakov systems in three dimensions. *Journal of Physics A: Mathematical and General, 26*, 4035–4046.
35. Gradshteyn, I. S., & Ryzhik, I. M. (1994). *Table of integrals, series and products* (5th ed.) A. Jeffrey Editor, Academic Press.
36. Fermi, E. (1950). High energy nuclear events. *Progress of Theoretical Physics (Japan), 5*, 570–583.
37. Mac Cullagh, J. (1831). On the intensity of light when the vibrations are elliptical. *Edinburgh Journal of Science*, 86–88; reprinted in *The Collected Works of James Mac Cullagh*, pp. 14–16.

Chapter 4
Motion on Spatial Multifractal Manifolds

4.1 Arguments

A complex system [1] is a very favorable medium for the appearance of instabilities. These instabilities imply both chaos through different routes (intermittencies, quasiperiodic, cascade of period-doubling bifurcations, subharmonic bifurcations, torus breakdown [2]) and self-construction through the generation of complex structures (double layers, multiple double layers, etc. [3, 4]). According to classical concepts, all theoretical complex system models (fluid models, kinetic models, etc.) assume that the dynamics of complex system structural units occur on continuous and differentiable curves so that they can be described in terms of continuous but differentiable motion variables (energy, momentum, density, etc.) [5, 6]. These motion variables are exclusively dependent on the spatial coordinates and time. In reality, complex system dynamics prove to be much more complex, and the above simplifications cannot be expected to explain all of the aspects of complex system dynamics. However, this situation can still be standardized if we consider that the complexity of complex system interaction processes imposes different temporal resolution scales, while complex evolution system patterns lead to different degrees of freedom [1, 7].

According to the abovementioned arguments, complex system dynamics can be explained by the assumption that the motions of complex system structural units occur on continuous but nondifferentiable curves (fractal curves) since, according to previous procedures [7–9], only a fractal curve is dependent on the scale resolution. Moreover, according to the methodology [2], the dynamics of special topologies that can implement evolution patterns in complex systems can lead to various degrees of freedom. Such an assumption can be sustained by a typical example related to the collision processes in complex systems: between two successive collisions, the trajectory of the complex system structural units is a straight line that becomes nondifferentiable at the impact point. Considering that all the collision impact points form an uncountable set of points, the trajectories of the complex system structural units become continuous but nondifferentiable curves.

M. Agop and S. A. Irimiciuc, *Multifractal Theory of Motion*,
https://doi.org/10.1007/978-981-97-6410-5_4

Since nondifferentiability (fractality) is a fundamental property of complex system dynamics, it seems necessary to construct corresponding nondifferentiable complex system physics. We assume that the complexity of interactions in complex system dynamics is replaced by nondifferentiability (fractality). This topic (fractal motion) was systematically developed using either scale relativity theory (SRT) [8, 9] or extended scale relativity theory (ESRT), i.e., scale relativity theory in an arbitrary constant fractal dimension [10–14].

Therefore, as described above, the Euclidean dynamics of a complex system with external constraints are replaced with the fractal dynamics of a complex system free of any external constraints. Practically, motion with constraints in Euclidean space, i.e., on continuous and differentiable curves, is replaced by motion free of constraints in fractal space, i.e., on continuous but nondifferentiable curves. To do this, we must admit the correspondence between the interaction processes and the "fractality" of the motion trajectories.

In such a context, for time scales large with respect to the inverse of the maximum Lyapunov exponent [2, 7], deterministic trajectories can be replaced by families of potential trajectories, and the concept of defined positions can be replaced by that of probability densities. Then, through the probability density, one can establish a relation between the fractal theory of motion given either by SRT or by ESRT and quantum measure theory [15, 16]. We note that in the following, the spatial coordinates are fractals, while the temporal coordinates are noted.

4.2 Nondifferentiability and Its Consequences

Assuming that the motions of the complex system structural units occur on continuous but nondifferentiable curves (fractal curves), the following consequences of the complex system dynamics emerge [8, 9]:

(i) Any continuous but nondifferentiable curve of the complex system structural units (fractal curve) is explicitly scale resolution δt dependent, i.e., its length tends to infinity when δt tends to zero.

We mention that, mathematically speaking, a curve is nondifferentiable if it satisfies the Lebesgue theorem, i.e., its length becomes infinite when the scale resolution goes to zero [7]. Consequently, in this limit, a curve is as zig-zagged as one can imagine. Thus, it exhibits self-similarity at every point, which can be translated into a property of holography (every part reflects the whole) [7].

Does not this happen in the complex system? This is why, from the smallest scale to the greatest-sized ones, the complex system is a whole!

(ii) The physics of complex system phenomena is related to the behavior of a set of functions during the zoom operation of the scale resolution δt. Then, through the substitution principle, δt will be identified with dt, i.e., $\delta t \equiv dt$, and consequently, it will be considered an independent variable. We reserve the notation δt for the usual time, as in Hamiltonian complex system dynamics;

(iii) The complex system dynamics are described through fractal variables, i.e., functions depending on both the spacetime coordinates and the scale resolution since the differential time reflection invariance of any dynamical variable is broken. Then, at any point on the fractal curve, two derivatives of the variable field $Q(t, dt)$ can be defined:

$$\frac{d_+Q(t,dt)}{dt} = \lim_{\Delta t \to 0+} \frac{Q(t+\Delta t, \Delta t) - Q(t\Delta t)}{\Delta t},$$
$$\frac{d_+Q(t,dt)}{dt} = \lim_{\Delta t \to 0-} \frac{Q(t, \Delta t) - Q(t - \Delta t, \Delta t).}{\Delta t} \tag{4.1}$$

The "+" sign corresponds to the forward process, while the "–" sign corresponds to the backward process;

(iv) The differential of the spatial coordinate field $dX^i(t, dt)$ is expressed as the sum of the two differentials, one of which is scale resolution independent (differential part$d_\pm x^i(t)$), and the other is scale resolution dependent (fractal part $d_\pm \xi^i(t)\big)$ [8, 9], i.e.,

$$d_\pm X^i(t, dt) = d_\pm x^i(t) + d_\pm \xi^i(t, dt) \tag{4.2}$$

(v) The nondifferentiable part of the spatial coordinate field satisfies the fractal equation [8, 9]:

$$d_\pm \xi^i(t, dt) = \lambda_\pm^i (dt)^{1/D_F} \tag{4.3}$$

where $\lambda_\pm^i$ are constant coefficients through which the fractalization type is specified and D_F defines the fractal dimension of the fractal curve.

In our opinion, complex system processes imply dynamic geodesics with various fractal dimensions. Specifically, for $D_F = 2$, quantum-type processes are generated. For $D_F < 2$, correlative-type processes are induced, while for $D_F > 2$, noncorrelative-type processes can be found (see for details [9]). We note that if the spatial manifold operates simultaneously with multiple fractal dimensions, D_F can be substituted with a singularity spectrum $f(\alpha)$ of order with $\alpha = \alpha(D_F)$ [7–9];

(vi) The differential time reflection invariance of any dynamical variable is recovered by combining the derivatives d_+/dt and d_-/dt in the nondifferentiable operator:

$$\frac{\overset{\prime}{d}}{dt} = \frac{1}{2}\left(\frac{d_+ + d_-}{dt}\right) - \frac{i}{2}\left(\frac{d_+ - d_-}{dt}\right) \tag{4.4}$$

This is a natural result of the complex prolongation procedure [17]. Applying the nondifferentiable operator to the spatial coordinate field yields the complex velocity field:

$$\hat{V}^i = \frac{d\hat{X}^i}{dt} = V_D^i - V_F^i \tag{4.5}$$

with

$$\begin{gathered}V_D^i = \frac{1}{2}\left(v_+^i + v_-^i\right), V_F^i = \frac{1}{2}\left(v_+^i - v_-^i\right)\\ v_+^i = \frac{d_+x^i + d_+\xi^i}{dt}, v_-^i = \frac{d_-x^i + d_-\xi^i}{dt}.\end{gathered} \tag{4.6}$$

The real part V_D^i is differentiable and scale resolution independent (differentiable velocity field), while the imaginary part V_F^i is nondifferentiable and scale resolution dependent (fractal velocity field);

(vii) In the absence of any external constraint, an infinite number of fractal curves (geodesics) can be found relating to any pair of points, and this is true on all scales. Then, in the fractal space, all complex system structural units are substituted with the geodesics themselves so that any external constraint is interpreted as a selection of geodesics by the measuring device. The infinity of geodesics in the bundle, their nondifferentiability, and the two values of the derivative imply a generalized statistical fluid such as a description (fractal fluid). Then, the average values of the fractal fluid variables must be considered in the previously mentioned sense, so the average $d_{\pm}X^i$ is [8, 9]

$$\left\langle d_{\pm}X^i\right\rangle \equiv d_{\pm}x^i \tag{4.7}$$

with

$$\left\langle d_{\pm}\xi^i\right\rangle = 0. \tag{4.8}$$

(viii) The complex system dynamics can be described through a covariant derivative, the explicit form of which is obtained as follows. Let us consider that the nondifferentiable curves are immersed in a three-dimensional space and that X^i are the spatial coordinate fields of a point on the nondifferentiable curve. We also consider a variable field $Q\left(X^i, t\right)$ and the following Taylor expansion up to the second order:

$$d_{\pm}Q\left(X^i, t\right) = \partial_t Q dt + \partial_i Q d_{\pm}X^i + \frac{1}{2}\partial_l\partial_k Q d_{\pm}X^l d_{\pm}X^k \tag{4.9}$$

These relations are valid at any point in space and are more valid for points X^i on the nondifferentiable curve that we selected in (4.9). From here, forward and backward values of (4.9) become

$$\langle d_{\pm}Q\rangle = \langle\partial_t Q dt\rangle + \left\langle\partial_i Q d_{\pm}X^i\right\rangle + \frac{1}{2}\left\langle\partial_l\partial_k Q d_{\pm}X^l d_{\pm}X^k\right\rangle \tag{4.10}$$

We suppose that the average values of the variable field Q and its derivatives coincide with each other and that the differentials $d_{\pm}X^{i}$ and dt are independent. Therefore, the average of their products coincides with the product of averages. Consequently, (4.10) becomes

$$d_{\pm}Q = \partial_t Q dt + \partial_i Q \langle d_{\pm}X^{i}\rangle + \frac{1}{2}\partial_l \partial_k Q \langle d_{\pm}X^{l} d_{\pm}X^{k}\rangle \tag{4.11}$$

Even if the average value of $d_{\pm}\xi^{i}$ is null, for the higher order of $d_{\pm}\xi^{i}$, the situation can still be different. Let us focus on the averages $d_{\pm}\xi^{l} d_{\pm}\xi^{k}$.. Using (4.3), we can write

$$\langle d_{\pm}\xi^{l} d_{\pm}\xi^{k}\rangle = \pm\lambda^{l}_{\pm}\lambda^{k}_{\pm}(dt)^{\left(\frac{2}{D_F}\right)-1}dt, \tag{4.12}$$

where the sign + corresponds to $dt > 0$ and the sign—corresponds to $dt < 0$.. Then, (4.11) takes the form:

$$\begin{aligned} d_{\pm}Q &= \partial_t Q dt + \partial_i Q \langle d_{\pm}X^{i}\rangle + \frac{1}{2}\partial_l \partial_k Q d_{\pm}x^{l} d_{\pm}x^{k} \\ &\pm \frac{1}{2}\partial_l \partial_k Q\left[\lambda^{l}_{\pm}\lambda^{k}_{\pm}(dt)^{(2/D_F)-1}dt\right]. \end{aligned} \tag{4.13}$$

If we divide by dt and neglect the terms that contain differential factors (for details, see the method from [10–14]), we obtain

$$\frac{d_{\pm}Q}{dt} = \partial_t Q + v^{i}_{\pm}\partial_i Q \pm \frac{1}{2}\lambda^{l}_{\pm}\lambda^{k}_{\pm}(dt)^{(2/D_F)-1}\partial_l \partial_k Q \tag{4.14}$$

These relations also allow us to define the coral operators:

$$\frac{d_{\pm}}{dt} = \partial_t + v^{i}_{\pm}\partial_i \pm \frac{1}{2}\lambda^{l}_{\pm}\lambda^{k}_{\pm}(dt)^{\left(\frac{2}{D_F}\right)-1}\partial_l \partial_k \tag{4.15}$$

Under these circumstances, taking into account (4.4), (4.5) and (4.15), let us calculate $\acute{d}/dt$. It results:

$$\frac{\acute{d}Q}{dt} = \partial_t Q + \acute{V}^{i}\partial_i Q + \frac{1}{4}(dt)^{(2/D_F)-1}D^{lk}\partial_l \partial_k Q, \tag{4.16}$$

Where

$$\begin{aligned} D^{lk} &= d^{lk} - i\overleftarrow{d}^{lk} \\ d^{lk} &= \lambda^{l}_{+}\lambda^{k}_{+} - \lambda^{l}_{-}\lambda^{k}_{-},\ \overleftarrow{d}^{lk} = \lambda^{l}_{+}\lambda^{k}_{+} + \lambda^{l}_{-}\lambda^{k}_{-}. \end{aligned} \tag{4.17}$$

Relation (4.16) also allows us to define the covariant derivative:

$$\frac{\hat{d}}{dt} = \partial_t + \hat{V}^i \partial_i + \frac{1}{4}(dt)^{(2/D_F)-1} D^{lk} \partial_l \partial_k. \tag{4.18}$$

4.3 Complex System Geodesics

Let us now consider the principle of scale covariance (the complex system physics laws are invariant with respect to scale transformations) and postulate that the passage from classical complex system physics to nondifferentiable (fractal) complex system physics can be implemented by replacing the standard time derivative d/dt with a nondifferentiable operator $\hat{d}/dt$. Thus, this operator plays the role of the covariant derivative; namely, it is used to write the fundamental equations of complex system dynamics in the same form as in the classic (differentiable) case. Under these conditions, applying the operator (4.18) to the complex velocity field (4.5), in the absence of any external constraint, the complex system geodesics have the following form:

$$\frac{\hat{d}\hat{V}^i}{dt} = \partial_t \hat{V}^i + \hat{V}^l \partial_l \hat{V}^i + \frac{1}{4}(dt)^{(2/D_F)-1} D^{lk} \partial_l \partial_k \hat{V}^i = 0 \tag{4.19}$$

This means that the local acceleration $\partial_t \hat{V}^i$, convection $\hat{V}^l \partial_l \hat{V}^i$ and dissipation $D^{lk} \partial_l \partial_k \hat{V}^i$ are balanced at any point on the nondifferentiable curve. Moreover, the presence of the viscosity-type complex coefficient $\frac{1}{4}(dt)^{(2/D_F)-1} D^{lk}$ indicates that the complex system is a rheological medium, so it has memory, as a datum, due to its own structure.

If fractalization is achieved by Markov-type stochastic processes, which involve Lévy-type movements of complex system structural units [7–9], then

$$\lambda_+^i \lambda_+^l = \lambda_-^i \lambda_-^l = 2\lambda\delta^{il}, \tag{4.20}$$

where δ^{il} is Kronecker's symbol, with the property

$$\delta^{il} = \begin{cases} 1, & i = l \\ 0, & i \neq l \end{cases}.$$

Under these conditions, the equation of complex system geodesics takes the following simple form:

$$\frac{\hat{d}\hat{V}^i}{dt} = \partial_t \hat{V}^i + \hat{V}^l \partial_l \hat{V}^i - i\lambda(dt)^{(2/D_F)-1}\partial^l \partial_l \hat{V}^i = 0 \tag{4.21}$$

Moreover, by separating the motions at differential and fractal scale resolutions,

$$\begin{aligned} \frac{\hat{d}V_D^i}{dt} &= \partial_t V_D^i + V_D^l \partial_l V_D^i - \left[V_F^l - \lambda(dt)^{(2/D_F)-1}\partial^l\right]\partial_l V_F^i = 0, \\ \frac{\hat{d}V_F^i}{dt} &= \partial_t V_F^i + V_D^l \partial_l V_F^i - \left[V_F^l - \lambda(dt)^{(2/D_F)-1}\partial^l\right]\partial_l V_D^i = 0. \end{aligned} \tag{4.22}$$

4.4 Geodesics in the Schrödinger-Type Scenario

For irrotational motions

$$\varepsilon_{ikl}\partial^k \hat{V}^l = 0, \tag{4.23}$$

where ε_{ikl} is the Levi–Civita pseudo tensor. Then, we choose $\hat{V}^l$ in the form [10–14]:

$$\hat{V}^i = -2i\lambda(dt)^{\left(\frac{2}{D_F}\right)-1}\partial^i \ln \Psi, \tag{4.24}$$

where for the moment, $\ln \Psi$ defines the scalar potential of the complex velocity field. Substituting (4.24) in (4.21), we obtain

$$\begin{aligned} \frac{\hat{d}\hat{V}^l}{dt} = &-2i\lambda(dt)^{(2/D_F)-1}\{\partial_t \partial^i \ln\Psi \\ &- i\left[2\lambda(dt)^{(2/D_F)-1}\left(\partial^l \ln \Psi \partial_l\right)\partial^i \ln \Psi \right. \\ &\left. + \lambda(dt)^{(2/D_F)-1}\partial^l \partial_l \partial^i \ln\Psi\right]\} = 0. \end{aligned} \tag{4.25}$$

Using the identities

$$\partial^l \partial_l \ln \Psi + \partial_i \ln \Psi \partial^i \ln \Psi = \frac{\partial_l \partial^l \Psi}{\Psi} \tag{4.26}$$

$$\partial^i\left(\frac{\partial^l \partial_l \Psi}{\Psi}\right) = 2\left(\partial^l \ln \Psi \partial_l\right)\partial^i \ln \Psi + \partial^l \partial_l \partial^i \ln \Psi \tag{4.27}$$

Equation (4.25) becomes

$$\frac{d\hat{V}^i}{dt} = -2i\lambda(dt)^{(2/D_F)-1}\partial^i\left[\partial_t \ln\psi - 2i\lambda(dt)^{(2/D_F)-1}\frac{\partial^l\partial_l\psi}{\psi}\right] = 0 \tag{4.28}$$

This equation can be integrated up to an arbitrary phase factor, which may be set to zero by a suitable choice of phase of Ψ and this yield: (the Schrödinger type scinario)

$$\lambda^2(dt)^{(4/D_F)-2}\partial^l\partial_l\Psi + i\lambda(dt)^{(2/D_F)-1}\partial_t\Psi = 0 \tag{4.29}$$

For motions on Peano-type curves, $D_F = 2$, at the Compton scale, $\lambda = \hbar/2m_0$, with $\hbar$, the reduced Planck's constant and m_0 the rest mass of the structural units, the relation (5.28) becomes the usual Schrödinger equation. From here, quantum measurement theory [15, 16] is developed. Equation (4.29) is a Schrödinger-type equation (complex system geodesics in the Schrödinger representation). We note that the presence of an external scalar potential U modifies Eq. (4.29) in the form:

$$\lambda^2(dt)^{(4/D_F)-2}\partial^l\partial_l\Psi + i\lambda(dt)^{(2/D_F)-1}\partial_t\Psi - \frac{U}{2}\Psi = 0 \tag{4.30}$$

Relation (4.30) corresponds to the equation of motion for the one-body problem in the Schrödinger-type representation of the nondifferentiable model (for other details, see [17]). The standard equation of motion for the "one body problem" in scale relativity theory [8, 9] is as follows

$$D^2\partial^l\partial_l\Psi + iD\partial_t\Psi - \frac{U}{2}\Psi = 0 \tag{4.31}$$

results from (4.30) for movements of the complex system structural units on Peano-type curves, $D_F = 2$ and the correspondence $\lambda \equiv D$, where D is the coefficient of the fractal nonfractal transition from scale relativity theory.

4.5 Geodesics in the Madelung-Type Scenario

If $\Psi = \sqrt{\rho}\exp(iS)$ with $\sqrt{\rho}$ is the amplitude and S as the phase of Ψ, the complex velocity field (4.24) takes the following explicit form:

$$\begin{aligned}
\hat{V}^i &= -2i\lambda(dt)^{(2/D_F)-1}\partial^i \ln\Psi, \\
V_D^i &= 2\lambda(dt)^{(2/D_F)-1}\partial^i S, \\
V_F^i &= \lambda(dt)^{(2/D_F)-1}\partial^i \ln\rho.
\end{aligned} \tag{4.32}$$

Substituting (4.31) in (4.21) and separating the real and imaginary parts up to an arbitrary phase factor, which may be set to zero by a suitable choice of the phase of

Ψ, we obtain the Madelung-type scenario:

$$\begin{aligned} \partial_t V_D^i + \left(V_D^l \partial_l\right) V_D^i &= -\partial^i(Q) \\ \partial_t \rho + \partial^i\left(\rho V_D^i\right) &= 0, \end{aligned} \tag{4.33}$$

where Q is the specific nondifferentiable (fractal) potential:

$$Q = -2\lambda^2 (dt)^{(4/D_F)-2} \frac{\partial^l \partial_l \sqrt{\rho}}{\sqrt{\rho}} = -\frac{V_F^l V_{Fl}}{2} - \lambda (dt)^{(2/D_F)-1} \partial_i V_F^i. \tag{4.34}$$

We note that the presence of an external scalar potential U modifies Eq. (4.33) in the form:

$$\partial_t V_D^i + \left(V_D^l \partial_l\right) V_D^i = -\partial^i(Q + U) \tag{4.35}$$

Equation (4.33) represents the specific momentum conservation law, while Eq. (4.34) represents the state density conservation law. Equations (4.33) and (4.34) with (4.35) define the fractal hydrodynamic model and imply the following:

(i) Any complex system structural unit is in permanent interaction with a fractal medium through a specific nondifferentiable potential (4.35);
(ii) The complex system can be identified with a fractal fluid (nondifferentiable fluid), the dynamics of which are described by the fractal hydrodynamic model;
(iii) The fractal velocity field V_F^i does not represent actual motion but contributes to the transfer of the specific momentum and to the energy focus. This may be clearly observed from the absence of V_F^i from the state density conservation law and from its role in the variational principle [9];
(iv) Any interpretation of the specific fractal potential should take cognizance of the "self" nature of the specific momentum transfer. While fractal energy is stored in the form of mass motion and fractal potential energy, some is available elsewhere, and only the total is conserved. The conservation of fractal energy and fractal momentum ensures fractal reversibility and the existence of fractal eigenstates but denies the Lévy motion fractal force of interaction with an external medium [7];
(v) The specific fractal potential Q induces the tensor (fractal stress tensor; for details, see [10]

$$\sigma'_{il} = -2\lambda^2 (dt)^{(4/D_F)-2} \rho \partial_i \partial_l \ln\rho \tag{4.36}$$

the divergence of which is equal to the usual force density associated with Q:

$$\partial^i \sigma'_{il} + \rho \partial_l Q = 0 \tag{4.37}$$

Under these conditions, the equations of fractal hydrodynamics become:

$$\begin{aligned} &\partial_t V_D^i + \left(V_D^l \partial_l\right) V_D^i = -\rho^{-1} \partial_l \sigma'^{il} + \partial^i U, \\ &\partial_t \rho + \partial_i\left(\rho V_D^i\right) = 0. \end{aligned} \tag{4.38}$$

(vi) Two types of fractal stationary states are distinguished:

(a) Fractal dynamic states. For $\partial_t = 0$ and $V_D^l \neq 0$, Eq. (4.39) gives

$$\begin{aligned} \tfrac{1}{2} V_D^l V_{Dl} + U + Q = E, \\ \rho V_D^i = \varepsilon^{ilk} \partial_l f_k. \end{aligned} \tag{4.39}$$

Therefore, the sum of the specific kinetic energy $V_D^l V_{Dl}/2$, external specific scalar potential U, and specific fractal potential Q is invariant, i.e., equal to the integration constant $E \neq E\left(x^l\right)$ (see the first Eq. (4.40)).

$E \equiv \langle E \rangle$ represents the total specific fractal energy of the fractal dynamic complex system. The fractal state density current, ρV_D^l, has no sources (see the second Eq. (4.40)), i.e., its fractal streamlines are closed.

(b) Fractal static states. For $\partial_t = 0$ and $V_D^l = 0$, Eq. (4.39) gives

$$U + Q = E \tag{4.40}$$

The sum of the external specific scalar potential, U, and the specific fractal potential, Q, is invariant, i.e., and is equal to the integration constant $E \neq E\left(x^l\right)$ (see Eq. (4.41)).

$E \equiv \langle E \rangle$ represents the total specific fractal energy of the fractal static complex system. The fractal state density conservation law is identically satisfied.

(vii) Let us define a logarithmic function

$$S\left(x^l, t\right) = \ln \rho\left(x^l, t\right), \tag{4.41}$$

which will be called the later nondifferentiable entropy. The nondifferentiable entropy evaluates the disorder degree of a complex system (for details, see [18, 19]).

In contrast, for the function:

$$I\left(x^l, t\right) = -\ln \rho\left(x^l, t\right). \tag{4.42}$$

That is, the negative nondifferentiable entropy characterizes the order degree of a complex system. This information will be called nondifferentiable information later. In any of the abovementioned alternatives, the dynamics, either of the nondifferentiable entropy or of the nondifferentiable information, are controlled by the specific fractal potential through the following relations:

$$Q = -\lambda^2 (dt)^{\left(\frac{4}{D_F}\right)-2} \left(\frac{1}{2} \partial_i S \partial^i S + \partial_i \partial^i S \right),$$

respectively,

$$Q = -\lambda^2 (dt)^{(4/D_F)-2} \left(\frac{1}{2} \partial_i I \partial^i I - \partial_i \partial^i I \right) \tag{4.43}$$

Moreover, by derivation, Eq. (4.43) has the following form:

$$F^i = -\partial^i Q = \lambda^2 (dt)^{\left(\frac{4}{D_F}\right)-2} \left(\frac{1}{2} \partial^i \partial_l I \partial^l I - \partial^i \partial_l \partial^l I \right), \tag{4.44}$$

which specifies that the variations in the nondifferentiable information induce force fields. These fields of forces depend not only on the type of interaction personalized by the constant of structures on the transition fractal-nonfractal through the constant λ but also on the fractalization "degree" given by the fractal dimension D_F.

4.6 Correlations Between the Schrödinger and Modeling Scenarios

The solution of the one-dimensional Schrödinger equation of fractal type (according to (4.29)) is

$$\mu_s^2 \partial_{l_s} \partial_s \Psi(x_s, t_s) + i\mu_s \partial_{t_s} \Psi(x_s, t_s) = 0, \tag{4.45}$$

Which can be written in the form

$$\Psi(x_s, t_s) = \frac{1}{\sqrt{t_s}} \exp\left(i \frac{x_s^2}{4\mu_s t_s} \right), \ \mu_s = \lambda_s [(dt)]^{\frac{2}{D_f}-1}, \ \mathrm{s} = 1, 2, 3 \tag{4.46}$$

and is defined, of course, up to an arbitrary multiplicative constant. In the above relations and in the following ones, the indexation with "s" defines the variables and parameters of dynamics on an anisotropic manifold, *dt* is the scale resolution and D_F is the fractal dimension.

As such, the general solution of Eq. (4.46) can be written as a linear superposition of the following form:

$$\Psi(x_s, t_s) = \frac{1}{\sqrt{t_s}} \int_{-\infty}^{+\infty} u(y_s) \exp\left[i \frac{(x_s - y_s)^2}{4\mu_s t_s} \right] dy_s \tag{4.47}$$

Now, if $u(y_s)$ is an Airy function of fractal type, then $\Psi(x_s, t_s)$ retains this property, in the sense that its amplitude is an Airy function of fractal type. Indeed, in this case, there will be

$$u(y_s) \equiv \mathrm{Ai}(y_s) = \frac{1}{2\pi} \int_{-\infty}^{+\infty} \exp\left[i\left(\frac{\omega_s^3}{3} + \omega_s y_s\right)\right] d\omega_s \tag{4.48}$$

in such a way that the state function (4.48) will be written in the following form:

$$\Psi(x_s, t_s) = \frac{1}{2\pi\sqrt{t_{u,s}}} \int_{-\infty}^{+\infty} \exp\left\{i\left[\frac{\omega_s^3}{3} + \omega_s y_s + \frac{(x_s - y_s)^2}{4\mu_s t_s}\right]\right\} dy_s \omega_s. \tag{4.49}$$

If, at first, the integration is carried out after y_s, up to a multiplicative constant, the results are

$$\Psi(x_s, t_s) = \frac{1}{2\pi} \int_{-\infty}^{+\infty} \exp\left[i\left(\frac{\omega_s^3}{3} + \omega x_s - \mu_s t_s \omega_s^2\right)\right] d\omega_s. \tag{4.50}$$

The final result is obtained based on a special relation developed in [20]:

$$\Psi(x_s, t_s) = \left[Ai\left(k_s x_s - v_s^2 t_s^2\right)\right] \exp\left[i v_s t_s \left(k_s x_s - \frac{2}{3} v_s^2 t_s^2\right)\right] \tag{4.51}$$

with

$$v_s = k_s^2 \mu_s. \tag{4.52}$$

Under these conditions, if Ψ is chosen in the form:

$$\Psi(x_s, t_s) = A(x_s, t_s) \exp[i\phi(x_s, t_s)], \tag{4.53}$$

where $A(x_s, t_s)$ is an amplitude and $\Phi(x_s, t_s)$ is a phase. By identifying the amplitude and the phase in (4.52), the following equation is obtained:

$$A(x_s, t_s) = Ai\left(k_s x_s - v_s^2 t_s^2\right), \tag{4.54}$$

$$\phi(x_s, t_s) = v_s t_s \left(k_s x_s - \tfrac{2}{3} v_s^2 t_s^2\right). \tag{4.55}$$

Considering the asymptotic behavior of the function $Ai(z)$ in its general form:

$$\mathrm{Ai}(z_S) \sim \begin{cases} \frac{1}{2\pi^{1/2}} z_S^{-1/4} \exp\left(-\frac{2}{3} z_S^{3/2}\right), & z_S \to +\infty \\ \frac{1}{\pi^{1/2}} |z_S|^{-1/4} \sin\left(\frac{2}{3}|z_S|^{3/2} + \frac{\pi}{4}\right), & z_S \to -\infty \end{cases} \tag{4.56}$$

the state (4.54) with (4.55), (4.56) function in the asymptotic limit $\Psi \to \Psi_A$ becomes

$$\Psi_A \sim \begin{cases} \frac{1}{2\pi^{1/2}} \left(k_s x_s - v_s^2 t_s^2\right)^{-1/4} \exp\left[-\frac{2}{3}\left(k_s x_s - v_s^2 t_s^2\right)^{3/2} + i v_s t_s \left(k_s x_s - \frac{2}{3} v_s^2 t_s^2\right)\right] \\ \frac{1}{\pi^{1/2}} \left|k_s x_s - v_s^2 t_s^2\right|^{-1/4} \sin\left[\frac{2}{3}\left|k_s x_s - v_s^2 t_s^2\right|^{3/2} + \frac{\pi}{4}\right] \exp\left[i v_s t_s \left(k_s x_s - \frac{2}{3} v_s^2 t_s^2\right)\right] \end{cases} \tag{4.57}$$

Figure 4.1 shows the 3D and contour plot representation of the wave function as defined in (4.54). The wave function can be influenced by a wide series of factors, including time. For relatively low values of the control parameter (t, v, k), the function closely follows the Airy-type representation, which is dominant in (4.54). When the system evolves at a longer moment of type, we observe self-modulation of the system and a change in frequency as both time and spatial coordinates are varied. The modulation appears on the temporal axis and defined a complex behavior on the spatial coordinate.

This means that by simply selecting the appropriate scale (defined by z, t, v, k), one can better investigate the particular dynamics of the system. Self-modulation is better when the properties of the wave are modified (through k), where it is possible to see that for $k = 6$, $v = 3$, and $t = 1$, the wave function defines a wave-like structure in the spatial coordinates of the scales, while in time, it complicates features with multiple oscillation frequencies. This means that by tailoring the resolution scale at which a system is investigated, it is possible to transition from a time- or space-modulated structure that characterizes particular phenomena. Further understanding of the space–time modulation of the wave function could become important when investigating transient phenomena such as laser-produced plasma or complex fluid flows, where there are often reports of temporal analysis for a fixed spatial volume or spatial analysis for a fixed moment of time [21–24].

By substituting (4.54) in (4.46) via direct calculation, the following relation is checked:

$$\begin{aligned} i\partial_{t_s}\Psi + \mu_s \partial_{l_s}\partial^{l_s}\psi = &-\left[\partial_{t_s}\phi + \mu_s\left(\partial_{l_s}\phi\right)^2 - \mu_s \frac{\partial_{l_s}\partial^{l_s} A}{A}\right] \\ &+ \frac{i}{2A^2}\left[\partial_{t_s} A^2 + 2\mu_s \partial_{l_s}\left(A^2 \partial^{l_s}\phi\right)\right] \end{aligned} \tag{4.58}$$

Now, the "specific constraints" necessary for Ψ to be a solution of the nonstationary differential Eq. (4.59) will be reducible to the differential equations:

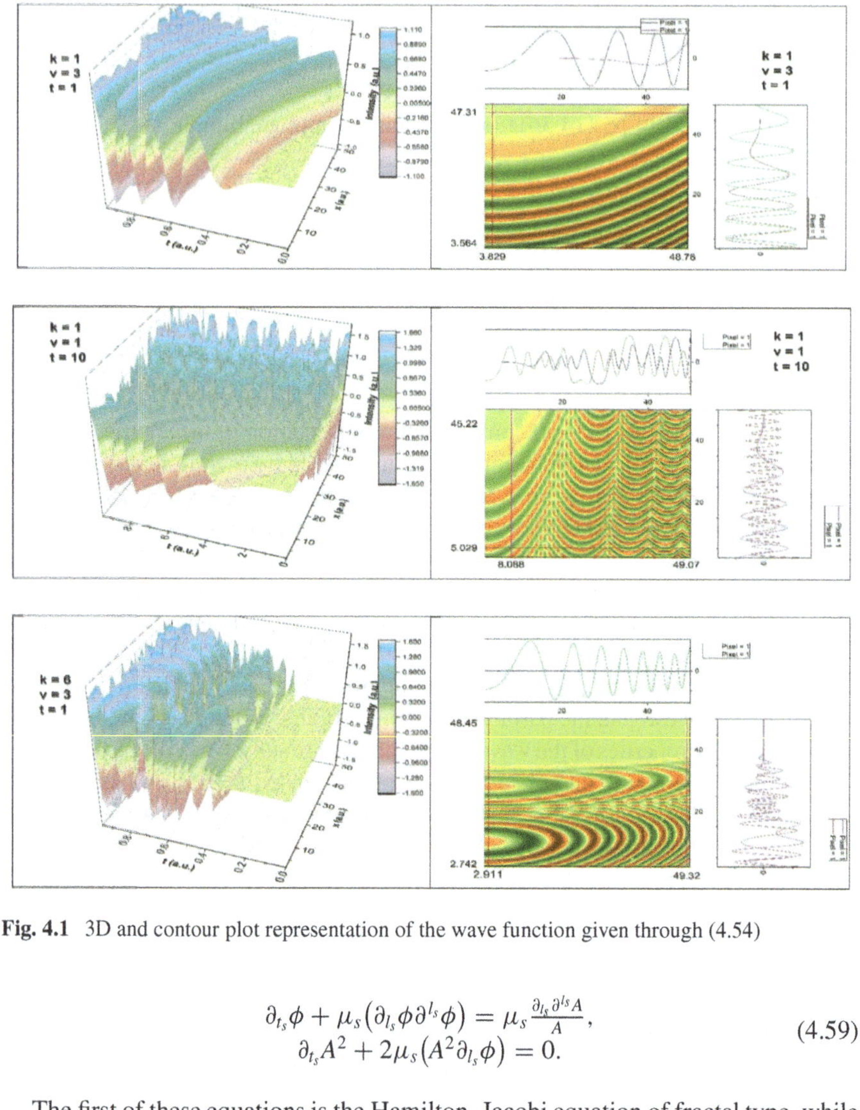

Fig. 4.1 3D and contour plot representation of the wave function given through (4.54)

$$\begin{gathered}\partial_{t_s}\phi+\mu_s\left(\partial_{l_s}\phi\partial^{l_s}\phi\right)=\mu_s\frac{\partial_{l_s}\partial^{l_s}A}{A},\\ \partial_{t_s}A^2+2\mu_s\left(A^2\partial_{l_s}\phi\right)=0.\end{gathered} \tag{4.59}$$

The first of these equations is the Hamilton–Jacobi equation of fractal type, while the second equation is the continuity equation of fractal type. The correspondence with the hydrodynamic model of fractal type pertaining to scale relativity [8, 9] becomes evident based on the following substitutions:

$$V_D^{i_s}=\mu_s\partial^{i_s}\phi,\ \rho=A^2, \tag{4.60}$$

where $V_D^{i_s}$ is the differential component of the velocity field and ρ is the density of states. The conservation law of the fractal type of the specific momentum can be found:

$$\partial_{t_s} V_D^{i_s} + V_D^{l_s} \partial_{l_s} V_D^{i_s} = -\partial^{i_s} Q \tag{4.61}$$

and, respectively, the conservation law of the density of states of fractal type:

$$\partial_{t_s} \rho + \partial^{l_s} \left(\rho V_D^{l_s} \right) = 0 \tag{4.62}$$

The specific potential of fractal type

$$Q = -\mu_s^2 \frac{\partial_{l_s} \partial^{l_s} \sqrt{\rho}}{\sqrt{\rho}} \tag{4.63}$$

through the induced specific force of fractal type:

$$f^{i_s} = -\partial^{i_s} Q = -\mu_s^2 \partial^{i_s} \left(\frac{\partial_{l_s} \partial^{l_s} \sqrt{\rho}}{\sqrt{\rho}} \right) \tag{4.64}$$

becomes a measure of the fractal degree pertaining to the motion curves. In such a motion, the "specific constraints" (4.60) are also checked in detail, with a specific potential of a fractal type

$$Q(x_s, t_s) = \mu_s v_s \left(k_s x_s - v_s^2 t_s^2 \right) \tag{4.65}$$

suppressing, in the scale relativity sense, dynamics with a constant force of fractal type.

As such, the nonstationary Schrödinger equation of fractal type in the scale space, in its "universal" instance given through (4.46), can lead to a wide range of interpretations. The "fractal object"/particle is in a uniformly accelerated motion of a fractal type (see both the argument of the Airy function of fractal type and the expression of the specific potential of fractal type). This evidently holds with the condition of accepting the functionality of the de Broglie theory of fractal type, which is linked to "wave phenomena named fractal objects". The generation of probability densities is given by the square of the Airy function of fractal type. As this cannot be integrated into the entire real straight line, the state/wave package of fractal type can have a center that is achievable in the sense of de Broglie's theory of fractal type. Then, the fractal-type state/wave package represents an ensemble of fractal objects/particles that all have a uniform rectilinear motion, but each has a different velocity, and the argument of the fractal-type Airy function represents a caustic in the dynamics space (the envelope of the ensemble of geodesics that represents the corresponding trajectories). Such an interpretation is linked with the nature of the invention of the Airy function: the behavior of light in the proximity of caustics. The functionality of an equivalence principle of fractal type implies that the Airy state/wave package of fractal type will not scatter because it represents a fractal object/particle in an enclosure analogous to Einstein's Elevator, and the uniform field of gravitational forces is thus suppressed.

References

1. Mitchell, M. (2009). *Complexity: a guided tour*. Oxford University Press.
2. Cristescu, C. P. (2008). *Dinamici Neliniare şi Haos*. Fundamente Teoretice şi Aplicaţii (in Romanian), Romanian Academy Publishing House.
3. Dimitriu, D. G. (2013). Experimental and theoretical investigations of plasma multiple double layers. *Plasma Sources Science and Technology, 22*, 035007.
4. Agop, M., Dimitriu, D. G., Niculescu, O., et al. (2013). Experimental and theoretical evidence for chaotic dynamics of complex systems. *Physica Scripta, 87*, 045501.
5. Flake, G. W. (1998). *The computational beauty of nature*. MA, MIT Press.
6. Lifshit, E. M., & Landau, L. M. (1987). *Fluid mechanics*. Pergamon Press.
7. Mandelbrot, B. B. (1983). The fractal geometry of nature (Updated and augm. ed.). New York: W.F. Freeman.
8. Nottale, L. (1993). *Fractal space-time and microphysics: Towards a theory of scale relativity*. World Scientific.
9. Nottale, L. (2011). *Scale relativity and fractal space-time—A new approach to unifying relativity and quantum mechanics*. Imperial College Press.
10. Munceleanu, G. V., Paun, V. P., Casian-Botez, I., & Agop, M. (2011). The microscopic-macrosopic scale transformation through a chaos scenario in the fractal space-time theory. *International Journal of Bifurcation and Chaos, 21*(2), 603–618.
11. Agop, M., & Casian-Botez, I. (2015). Theoretical approach of differential-non-differential scale transition in nanostructures. *JCTN, 12*, 1–12.
12. Agop, M., Forna, N., & Casian-Botez, I. (2008). New theoretical approach of the physical processes in nanostructures. *Journal of Computational and Theoretical Nanoscience, 5*(4), 483–489.
13. Colotin, M., Pompilian, G. O., Nica, P., Gurlui, S., Paun, V., & Agop, M. (2009). Frac- tal transport phenomena through the scale relativity model. *Acta Physica Polonica A, 116*(2), 157–164.
14. Timofte, A., Casian, B. I., Scurtu, D., & Agop, M. (2011). Systems dynamics control through the fractal potential. *Acta Physica Polonica A, 119*(3), 304–311.
15. Sorkin, R. (1994). Quantum mechanics as quantum measure theory. *Modern Physics Letters A, 9*, 3119–3127.
16. Sorjin, R. (2007). Quantum mechanics without the wave function. *Journal of Physics A, 40*, 3207–3231.
17. Cresson, J., & Adda, F. B. (2004). Quantum derivatives and the Schrodinger equation. *Chaos, Solitons and Fractals, 19*(5), 1323–1334.
18. Agop, M., Gavrilu¸t, A., Crumpei, G., & Doroftei, B. (2014). Informational non-differentiable entropy and uncertainty relations in complex systems. *Entropy, 16*(11), 6042–6058.
19. Agop, M., Gavrilut, A., Stefan, G., & Doroftei, B. (2015). Implications of non-differentiable entropy on a space-time manifold. *Entropy, 17*(4), 2184–2197.
20. Gradshteyn, I. S., & Ryzhik, I. M. (1994). *Table of integrals, series and products* (5th ed.). A. Jeffrey Editor, Academic Press.
21. Irimiciuc, S. A., Chertopalov, S., Lancok, J., & Craciun, V. (2021). Langmuir probe technique for plasma characterization during pulsed laser deposition process. *Coatings, 11*(7), 762.
22. Schrittwieser, R. W., Ionita, C., Teodorescu-Soare, C. T., Vasilovici, O., Gurlui, S., Irimiciuc, S. A., & Dimitriu, D. G. (2017). Spectral and electrical diagnosis of complex space-charge structures excited by a spherical grid cathode with orifice. *Physica Scripta, 92*(4), 044001.
23. Irimiciuc, S. A., Gurlui, S., & Agop, M. (2019). Particle distribution in transient plasmas generated by ns-laser ablation on ternary metallic alloys. *Applied Physics B, 125*(10).
24. Agop, M., Cimpoesu, N., Gurlui, S., & Irimiciuc, S. A. (2020). Investigations of transient plasma generated by laser ablation of hydroxyapatite during the pulsed laser deposition process. *Symmetry, 12*(1), 132.

Chapter 5
Motions for Space–Time Multifractal Manifolds

5.1 Arguments

By analyzing the motion of a particle on a fractal curve [1–4], we observe a large discrepancy between the spatial coordinates and the temporal coordinates (considered an affine parameter of the motion curve), the latter being nonfractal. This discrepancy also has an immediate abnormal consequence: the particle travels on an infinite length curve in a finite time span, and thus, it has an infinite velocity. To eliminate this contradiction, we assume that the temporal coordinate of the fractal curve is also fractal. Thus, most elements of the nonrelativistic approach of scale relativity theory with arbitrary constant fractal dimensions, as described in [5–11], remain valid, but the time differential element *dt* is now replaced by the proper time differential element $d\tau$. In this way, not only the space but also the entire space–time continuum is considered to be nondifferentiable and, therefore, fractal. In this framework, in the present chapter, we extend the results from [12, 20].

5.2 Nondifferentiability and Its Consequences

Let us suppose that in a Minkowski-type space–time, the motions of structural units (complex system components [13–18]) take place on continuous but nondifferentiable curves (in particular, fractal curves). The nondifferentiability of motion curves implies the following:

(i) Any continuous but nondifferentiable curve is explicitly scale dependent (which will be referred to as $\delta\tau$). In other words, its length tends to infinity when its proper time interval, $\Delta\tau$, tends to zero. Then, a continuous but nondifferentiable space–time fractal is used in Mandelbrot's sense [4];

M. Agop and S. A. Irimiciuc, *Multifractal Theory of Motion*,
https://doi.org/10.1007/978-981-97-6410-5_5

(ii) The differential proper time reflection invariance of any field variable is broken. For example, the proper time derivative of the 4-coordinate field X^{μ} can be written as follows:

$$\left[\frac{dX^{\mu}}{d\tau}\right]_{+} = \lim_{\Delta\tau \to 0_{+}} \frac{X^{\mu}(\tau + \Delta\tau) - X^{\mu}(\tau)}{\Delta\tau},$$
$$\left[\frac{dX^{\mu}}{d\tau}\right]_{-} = \lim_{\Delta\tau \to 0_{-}} \frac{X^{\mu}(\tau) - X^{\mu}(\tau - \Delta\tau)}{\Delta\tau}. \quad (5.1)$$

These relations are equivalent in the differentiable case, $\Delta\tau \to -\Delta\tau$. In the nondifferentiable case, the previous definitions fail since the limits $\Delta\tau \to 0_{\pm}$ are no longer defined. In the approach of the nondifferentiable model, physical phenomena are related to the behavior of the function during the "zoom" operation at the proper temporal resolution $\delta\tau$. Then, by means of the substitution principle, $\delta\tau$ will be identified with the differential element $d\tau$, i.e., $\delta\tau \equiv d\tau$, and will be considered an independent variable. Thus, every standard field $Q(\tau)$ is replaced by the nondifferentiable field $Q(\tau, d\tau)$, which is explicitly dependent on the proper temporal resolution interval whose derivative is undefined only within the unobservable limit, $\Delta\tau \to 0$. As a consequence, two derivatives of every nondifferentiable field are defined as explicit functions of τ and $d\tau$. For example, the two derivatives of the 4-coordinate field $X^{\mu}(\tau, \Delta\tau)$ take the following form:

$$\frac{d_{+}X^{\mu}}{d\tau} = \lim_{\Delta\tau \to 0_{+}} \frac{X^{\mu}(\tau + \Delta\tau, \Delta\tau) - X^{\mu}(\tau, \Delta\tau)}{\Delta\tau},$$
$$\frac{d_{-}X^{\mu}}{d\tau} = \lim_{\Delta\tau \to 0_{-}} \frac{X^{\mu}(\tau, \Delta\tau) - X^{\mu}(\tau - \Delta\tau, \Delta\tau)}{\Delta\tau}. \quad (5.2)$$

The + symbol corresponds to the forward physical process, and the—symbol corresponds to the backward process;

(iii) The differential of the 4-coordinate field $dX^{\mu}(\tau, \Delta\tau)$ can be expressed as the sum of two differentials, one not scales dependent (differentiable part $d_{\pm}x^{\mu}(\tau)$) and the other scale-dependent (nondifferentiable part $d_{\pm}\xi^{\mu}(\tau, d\tau)$), as follows:

$$d_{\pm}X^{\mu}(\tau, \Delta\tau) = d_{\pm}x^{\mu}(\tau) + d_{\pm}\xi^{\mu}(\tau, d\tau). \quad (5.3)$$

(iv) The nondifferentiable part of the 4-coordinate field satisfies the nondifferentiable equation:

$$d_{\pm}\xi^{\mu}(\tau, d\tau) = \lambda_{\pm}^{\mu}(d\tau)^{\frac{1}{D_F}}, \quad (5.4)$$

where $\lambda_{\pm}$ are constant coefficients, whose statistical significance is given in what follows and D_F is the fractal dimension of the motion curves. If the maniford is operating simultaneously with multiple fractal dimensions, then D_F can be substituted with a singularity spectrum f(α) of α order with $\alpha = \alpha(D_F)$;

(v) The differential proper time reflection invariance is recovered by combining the derivatives $\frac{d_+}{d\tau}$ and $\frac{d_-}{d\tau}$ in the nondifferentiable operator:

$$\frac{\hat{d}}{d\tau} = \frac{1}{2}\left(\frac{d_+ + d_-}{d\tau}\right) - \frac{i}{2}\left(\frac{d_+ - d_-}{d\tau}\right). \tag{5.5}$$

This specific procedure is called, according to [19], "extension in complex by differentiability". Applying the nondifferentiable operator to the 4-coordinate field yields the complex velocity:

$$\begin{aligned}\hat{V}^{\mu} &= \frac{\hat{d} X^{\mu}}{d\tau} = \frac{1}{2}\left(\frac{d_+X^{\mu} + d_-X^{\mu}}{d\tau}\right) - \frac{i}{2}\left(\frac{d_+X^{\mu} - d_-X^{\mu}}{d\tau}\right)\\ &= \frac{1}{2}\left(\frac{d_+x^{\mu} + d_-x^{\mu}}{d\tau} + \frac{d_+\xi^{\mu} + d_-\xi^{\mu}}{d\tau}\right)\\ &\quad - \frac{i}{2}\left(\frac{d_+x^{\mu} - d_-x^{\mu}}{d\tau} + \frac{d_+\xi^{\mu} - d_-\xi^{\mu}}{d\tau}\right) = V^{\mu} - iU^{\mu}\end{aligned} \tag{5.6}$$

with

$$\begin{aligned}V^{\mu} &= \frac{1}{2}(v_+^{\mu} + v_-^{\mu}),\ U^{\mu} = \frac{1}{2}(v_+^{\mu} - v_-^{\mu}),\\ v_+^{\mu} &= \frac{d_+x^{\mu} + d_+\xi^{\mu}}{d\tau},\ v_-^{\mu} = \frac{d_-x^{\mu} + d_-\xi^{\mu}}{d\tau}.\end{aligned} \tag{5.7}$$

The real part V^{μ} is differentiable and scale resolution independent, while the imaginary part U^{μ} is nondifferentiable and scale resolution dependent;

(vi) There can be an infinite number of nondifferentiable curves (geodesics) relating any pair of points. This is true at all scales. The structural unit is substituted with the geodesics themselves so that any measurement is interpreted as a selection of geodesics by the measuring device. The infinite geodesics in the bundle, their nondifferentiability, and the two values of the derivative imply a generalized statistical fluid, such as a description (nondifferentiable fluid). Then, the average values of the fluid variables must be considered in the previously mentioned sense, so the average of $d_{\pm}X^{\mu}$ is

$$\langle d_{\pm}X^{\mu}\rangle \equiv d_{\pm}x^{\mu} \tag{5.8}$$

with

$$\langle d_{\pm}\xi^{\mu}\rangle \equiv 0. \tag{5.9}$$

5.3 Geodesics

Let us now consider that the movement curves (continuous but nondifferentiable) are immersed in space time and that X^{μ} are the 4 coordinates of a point on the curve. We also consider a variable field $Q(X^{\mu}, \tau)$ and the following Taylor expansion up to the second order:

$$d_{\pm}Q(X^{\mu}, \tau) = \partial_{\tau}Qd\tau + \partial_{\mu}Qd_{\pm}X^{\mu} + \frac{1}{2}\partial_{\mu}\partial_{\nu}Qd_{\pm}X^{\mu}d_{\pm}X^{\nu}. \tag{5.10}$$

The relations (7.9) are valid at any point of the space–time manifold and more for the points "X^{μ}" on the nondifferentiable curve, which we have selected in relation to (7.9) From here, the forward and backward average values of (7.9) become

$$\langle d_{\pm}Q(X^{\mu}, \tau)\rangle = \langle \partial_{\tau}Qd\tau\rangle + \langle \partial_{\mu}Qd_{\pm}X^{\mu}\rangle + \frac{1}{2}\langle \partial_{\mu}\partial_{\nu}Qd_{\pm}X^{\mu}d_{\pm}X^{\nu}\rangle. \tag{5.11}$$

We make the following stipulations: the average values of the variable field $Q(X^{\mu}, \tau)$ and its derivatives coincide with each other, and the differentials $d_{\pm}X^{\mu}$ and $d\tau$ are independent. Therefore, the average of their products coincides with the product of their averages. Under these conditions, (7.11) takes the following form:

$$d_{\pm}Q(X^{\mu}, \tau) = \partial_{\tau}Qd\tau + \partial_{\mu}Q\langle d_{\pm}X^{\mu}\rangle + \frac{1}{2}\partial_{\mu}\partial_{\nu}Q\langle d_{\pm}X^{\mu}d_{\pm}X^{\nu}\rangle \tag{5.12}$$

or, using (5.3), (5.8), and (5.9):

$$d_{\pm}Q(X^{\mu}, \tau) = \partial_{\tau}Qd\tau + \partial_{\mu}Qd_{\pm}x^{\mu} + \frac{1}{2}\partial_{\mu}\partial_{\nu}Q(d_{\pm}x^{\mu}d_{\pm}x^{\nu} + \langle d_{\pm}\xi^{\mu}d_{\pm}\xi^{\nu}\rangle). \tag{5.13}$$

Even if the average value of the 4-nondifferentiable coordinate $d_{\pm}\xi^{\mu}$ is null, for the higher order of the 4-nondifferentiable coordinate average, the situation can be different. Now, we focus on the mean $\langle d_{\pm}\xi^{\mu}d_{\pm}\xi^{\nu}\rangle$.. Using (7.4), we can write

$$\langle d_{\pm}\xi^{\mu}d_{\pm}\xi^{\nu}\rangle = \pm\lambda_{\pm}^{\mu}\lambda_{\pm}^{\nu}(d\tau)^{\left(\frac{2}{D_F}\right)-1}d\tau \tag{5.14}$$

using the convention that the sign + corresponds to $d\tau > 0$, while the sign corresponds to $d\tau < 0$.

Then, (5.13) takes the form

$$\begin{aligned} d_{\pm}Q(X^{\mu}, \tau) &= \partial_{\tau}Qd\tau + \partial_{\mu}Qd_{\pm}x^{\mu} + \frac{1}{2}\partial_{\mu}\partial_{\nu}Qd_{\pm}x^{\mu}d_{\pm}x^{\nu} \\ &\pm \frac{1}{2}\partial_{\mu}\partial_{\nu}Q\lambda_{\pm}^{\mu}\lambda_{\pm}^{\nu}(d\tau)^{\left(\frac{2}{D_F}\right)-1}d\tau. \end{aligned} \tag{5.15}$$

If we divide by $d\tau$ and neglect the terms that contain differential factors, using the method from [5–11], we obtain

$$\frac{d_{\pm}Q(X^{\mu},\tau)}{d\tau}=\partial_{\tau}Q+v_{\pm}^{\mu}\partial_{\mu}Q\pm\frac{1}{2}\lambda_{\pm}^{\mu}\lambda_{\pm}^{\nu}(d\tau)^{\left(\frac{2}{D_F}\right)-1}\partial_{\mu}\partial_{\nu}Q. \tag{5.16}$$

These relations also allow us to define the operators:

$$\frac{d_{\pm}}{d\tau}=\partial_{\tau}+v_{\pm}^{\mu}\partial_{\mu}\pm\frac{1}{2}\lambda_{\pm}^{\mu}\lambda_{\pm}^{\nu}(d\tau)^{\left(\frac{2}{D_F}\right)-1}\partial_{\mu}\partial_{\nu}. \tag{5.17}$$

Under these circumstances, let us calculate $\frac{\hat{\partial}}{d\tau}$. Taking into account (5.5), (5.6) and (5.17), we obtain

$$\begin{aligned}
\frac{\hat{d}Q}{d\tau}&=\frac{1}{2}\left[\left(\frac{d_{+}Q+d_{-}Q}{d\tau}\right)-i\left(\frac{d_{+}Q-d_{-}Q}{d\tau}\right)\right]\\
&=\frac{1}{2}\Bigg\{\left[\partial_{\tau}Q+v_{+}^{\mu}\partial_{\mu}Q+\frac{1}{2}\lambda_{+}^{\mu}\lambda_{+}^{\nu}(d\tau)^{\left(\frac{2}{D_F}\right)-1}\partial_{\mu}\partial_{\nu}Q\right]\\
&\quad+\left[\partial_{\tau}Q+v_{-}^{\mu}\partial_{\mu}Q-\frac{1}{2}\lambda_{-}^{\mu}\lambda_{-}^{\nu}(d\tau)^{\left(\frac{2}{D_F}\right)-1}\partial_{\mu}\partial_{\nu}Q\right]\Bigg\}\\
&\quad-\frac{i}{2}\Bigg\{\left[\partial_{\tau}Q+v_{+}^{\mu}\partial_{\mu}Q+\frac{1}{2}\lambda_{+}^{\mu}\lambda_{+}^{\nu}(d\tau)^{\left(\frac{2}{D_F}\right)-1}\partial_{\mu}\partial_{\nu}Q\right]\\
&\quad-\left[\partial_{\tau}Q+v_{-}^{\mu}\partial_{\mu}Q-\frac{1}{2}\lambda_{-}^{\mu}\lambda_{-}^{\nu}(d\tau)^{\left(\frac{2}{D_F}\right)-1}\partial_{\mu}\partial_{\nu}Q\right]\Bigg\}\\
&=\partial_{\tau}Q+\left(\frac{v_{+}^{\mu}+v_{-}^{\mu}}{2}-i\frac{v_{+}^{\mu}-v_{-}^{\mu}}{2}\right)\partial_{\mu}Q\\
&\quad+\frac{1}{4}(d\tau)^{\left(\frac{2}{D_F}\right)-1}\left[\left(\lambda_{+}^{\mu}\lambda_{+}^{\nu}-\lambda_{-}^{\mu}\lambda_{-}^{\nu}\right)-i\left(\lambda_{+}^{\mu}\lambda_{+}^{\nu}+\lambda_{-}^{\mu}\lambda_{-}^{\nu}\right)\right]\partial_{\mu}\partial_{\nu}Q\\
&=\partial_{\tau}Q+\hat{V}^{\mu}\partial_{\mu}Q+\frac{1}{4}(d\tau)^{\left(\frac{2}{D_F}\right)-1}D^{\mu\nu}\partial_{\mu}\partial_{\nu}Q,
\end{aligned} \tag{5.18}$$

where

$$\begin{aligned}
&D^{\mu\nu}=d^{\mu\nu}-i\overleftarrow{d}^{\mu\nu}\\
&d^{\mu\nu}=\lambda_{+}^{\mu}\lambda_{+}^{\nu}-\lambda_{-}^{\mu}\lambda_{-}^{\nu},\ \overleftarrow{d}^{\mu\nu}=\lambda_{+}^{\mu}\lambda_{+}^{\nu}+\lambda_{-}^{\mu}\lambda_{-}^{\nu}.
\end{aligned} \tag{5.19}$$

This relation also allows us to define the nondifferentiable operator:

$$\frac{\hat{d}}{d\tau}=\partial_{\tau}+\hat{V}^{\mu}\partial_{\mu}+\frac{1}{4}(d\tau)^{\left(\frac{2}{D_F}\right)-1}D^{\mu\nu}\partial_{\mu}\partial_{\nu}. \tag{5.20}$$

If the nondifferentiability of the motion curve is realized through a Markov-type stochastic process [1, 2, 4]:

$$\lambda_+^{\mu}\lambda_+^{\nu} = \lambda_-^{\mu}\lambda_-^{\nu} = -\lambda\eta^{\mu\nu}, \tag{5.21}$$

where $\eta^{\mu\nu}$ is the Minkowski metric; then, the nondifferentiable operator takes the form:

$$\frac{\hat{d}}{d\tau} = \partial_\tau + \hat{V}^{\mu}\partial_\mu + i\frac{\lambda}{2}(d\tau)_\mu^{\left(\frac{2}{D_F}\right)-1}\partial_\mu\partial^\nu. \tag{5.22}$$

5.4 Geodesies in the Klein-Gordon-Type Scenario

We now apply the principle of scale covariance [1, 2] (physics laws are simultaneously invariant both with respect to the 4-coordinate transformations and with respect to scale transformations) and postulate that the passage from differentiable mechanics to the nondifferentiable mechanics, which is considered here, can be implemented by replacing the standard time derivative $d/d\tau$ with the nondifferentiable operator $\hat{d}/d\tau$. This operator plays the role of a "covariant derivative operator", namely, it is used to write the fundamental equations of dynamics in the same form as in the classical and differentiable cases. Thus, by applying the operator (5.20) to the complex velocity (5.6), the geodesic equation is obtained

$$\frac{\hat{d}\hat{V}^{\mu}}{d\tau} = \partial_\tau\hat{V}^{\mu} + \hat{V}^{\nu}\partial_\nu\hat{V}^{\mu} + \frac{1}{4}(d\tau)^{\left(\frac{2}{D_F}\right)-1}D^{\alpha\beta}\partial_\alpha\partial_\beta\hat{V}^{\mu} \equiv 0 \tag{5.23}$$

or, using (5.6), through separation of motions at various resolutions (separation of the real part from the imaginary one):

$$\begin{aligned}\frac{\hat{d}V^{\mu}}{d\tau} &= \partial_\tau V^{\mu} + V^{\nu}\partial_\nu V^{\mu} - U^{\nu}\partial_\nu U^{\mu} + \frac{1}{4}(d\tau)^{(2/D_F)-1}d^{\alpha\beta}\partial_\alpha\partial_\beta V^{\mu}\\ &\quad - \frac{1}{4}(d\tau)^{\left(\frac{2}{D_F}\right)-1}\overleftarrow{d}^{\alpha\beta}\partial_\alpha\partial_\beta U^{\mu} = 0\\ \frac{\hat{d}U^{\mu}}{d\tau} &= \partial_\tau U^{\mu} + V^{\nu}\partial_\nu U^{\mu} + U^{\nu}\partial_\nu V^{\mu} + \frac{1}{4}(d\tau)^{(2/D_F)-1}d^{\alpha\beta}\partial_\alpha\partial_\beta U^{\mu}\\ &\quad + \frac{1}{4}(d\tau)^{(2/D_F)-1}\overleftarrow{d}^{\alpha\beta}\partial_\alpha\partial_\beta V^{\mu} = 0.\end{aligned} \tag{5.24}$$

For motions on nondifferentiable curves through a Markov-type stochastic process [1, 2, 4], the geodesic equation takes the form:

$$\frac{d\hat{V}^{\mu}}{d\tau} = \partial_\tau \hat{V}^{\mu} + \hat{V}^{\nu}\partial_\nu \hat{V}^{\mu} + i\frac{\lambda}{2}(d\tau)^{\left(\frac{2}{D_F}\right)-1}\partial^\nu\partial_\nu \hat{V}^{\mu} = 0 \tag{5.25}$$

or, through scale resolution separation

$$\frac{\hat{d}V^{\mu}}{d\tau} = \partial_\tau V^{\mu} + V^{\nu}\partial_\nu V^{\mu} - \left(U^{\mu} - \frac{\lambda}{2}(d\tau)^{\left(\frac{2}{D_F}\right)-1}\partial^\nu\right)\partial_\nu U^{\mu} = 0,$$

$$\frac{\hat{d}U^{\mu}}{d\tau} = \partial_\tau U^{\mu} + V^{\nu}\partial_\nu U^{\mu} - \left(U^{\mu} - \frac{\lambda}{2}(d\tau)^{\left(\frac{2}{D_F}\right)-1}\partial^\nu\right)\partial_\nu V^{\mu} = 0. \tag{5.26}$$

Let us choose $\hat{V}^{\mu}$ in terms of the wave function Ψ:

$$\hat{V}_{\alpha} = i\lambda(d\tau)^{(2/D_F)-1}\partial_\alpha \ln\Psi. \tag{5.27}$$

Then, the geodesic Eq. (5.25) becomes

$$\begin{aligned}\frac{d\hat{V}_{\alpha}}{d\tau} &= i\lambda(d\tau)^{(2/D_F)-1}\partial_\tau\partial_\alpha \ln\Psi + \left[i\lambda(d\tau)^{(2/D_F)-1}\partial^\mu \ln\Psi\right.\\ &\quad \left.+ \frac{i\lambda}{2}(d\tau)^{(2/D_F)-1}\partial_\mu\right]\partial_\mu\partial_\alpha\left[i\lambda(d\tau)^{(2/D_F)-1}\ln\Psi\right] = 0.\end{aligned} \tag{5.28}$$

Since

$$\begin{aligned}\partial_\alpha\left(\partial_\mu \ln\Psi\,\partial^\mu \ln\Psi\right) &= 2\partial^\mu \ln\Psi\,\partial_\alpha\partial_\mu \ln\Psi\\ \partial_\alpha\partial_\mu\partial^\mu \ln\Psi &= \partial^\mu\partial_\mu\partial_\alpha \ln\Psi\\ \partial_\alpha\left(\partial_\mu \ln\Psi\,\partial^\mu \ln\Psi + \partial_\mu\partial^\mu \ln\Psi\right) &= \partial_\alpha\left(\frac{\partial_\mu\partial^\mu\Psi}{\Psi}\right),\end{aligned} \tag{5.29}$$

Equation (5.28) becomes

$$i\lambda(d\tau)^{\left(\frac{2}{D_F}\right)-1}\partial_\tau\partial_\alpha \ln\Psi + \lambda^2(d\tau)^{\left(\frac{4}{D_F}\right)-2}\partial_\alpha\left(\frac{\partial_\mu\partial^\mu\Psi}{\Psi}\right) = 0. \tag{5.30}$$

By integrating the above relation up to an arbitrary phase factor, which may be set constant by a suitable choice of the phase of Ψ, we obtain

$$\lambda^2(d\tau)^{\left(\frac{4}{D_F}\right)-2}\partial_\mu\partial^\mu\Psi + i\lambda(d\tau)^{\left(\frac{2}{D_F}\right)-1}\partial_\tau\Psi + \omega^2\Psi = 0, \tag{5.31}$$

where ω is an integration constant. Here, if the wave function is independent of τ, $\partial_\tau\Psi = 0$, the relation (5.31) becomes a nondifferentiable Klein–Gordon-type equation (Klein–Gordon-type scenario):

$$\partial_\mu \partial^\mu \Psi + \frac{1}{\overleftarrow{\Lambda}^2} \Psi = 0 \tag{5.32}$$

with

$$\overleftarrow{\Lambda} = \overleftarrow{\Lambda}_0 (d\tau)^{\left(\frac{2}{D_F}\right)-1}, \; \overleftarrow{\Lambda}_0 = \frac{\lambda}{\omega}. \tag{5.33}$$

According to (5.33), the scale resolution depends on the fundamental length $\overleftarrow{\Lambda}$ [1, 2, 4], where $\overleftarrow{\Lambda}_0$ is the fundamental unscaled length. For motions on the Peano curves, $D_F = 2$, at the Compton scale $\overleftarrow{\Lambda}_0 = \lambda/\overleftarrow{\omega} = \hbar/m_0 c, \lambda = \hbar/m_0, \overleftarrow{\omega} \equiv c$[1, 2], where $\hbar$ is the reduced Planck constant, m_0 is the rest mass of the particle and c is the vacuum velocity light (5.33), which takes the usual form of the Klein–Gordon equation [20]:

$$\partial_\mu \partial^\mu \Psi + (m_0 c/\hbar)^2 \Psi = 0.$$

5.5 Multifractal Force and Energy

Using the explicit form of the wave function, $\Psi = \sqrt{\rho} e^{iS}$, where $\sqrt{\rho}$ is an amplitude and S is a phase, the expression of U_α becomes

$$U_\alpha = -\lambda \partial_\alpha \ln \sqrt{\rho}. \tag{5.34}$$

Thus, it results:

$$\left[U_\mu - \frac{\lambda}{2}(d\tau)^{(2/D_F)-1}\partial_\mu\right]\partial^\mu U_\alpha = \lambda^2 (d\tau)^{(4/D_F)-2}\Big(\partial^\mu \ln\sqrt{\rho}\,\partial_\mu \partial_\alpha \ln\sqrt{\rho} + \frac{1}{2}\partial^\mu \partial_\mu \partial_\alpha \ln\sqrt{\rho}\Big) \tag{5.35}$$

Since the identities from (5.29) also work in variable $\ln\sqrt{\rho}$, Eq. (5.35)

$$\begin{aligned}
&\left[U_\mu - \frac{\lambda}{2}(d\tau)^{(2/D_F)-1}\partial_\mu\right]\partial^\mu U_\alpha \\
&= \frac{\lambda^2}{2}(d\tau)^{\left(\frac{4}{D_F}\right)-2}\partial_\alpha\left(\partial^\mu \ln\sqrt{\rho}\,\partial_\mu \ln\sqrt{\rho} + \partial^\mu \partial_\mu \ln\sqrt{\rho}\right) \\
&= \frac{1}{2}\lambda^2 (d\tau)^{\left(\frac{4}{D_F}\right)-2}\partial_\alpha\left(\frac{\partial^\mu \partial_\mu \sqrt{\rho}}{\sqrt{\rho}}\right),
\end{aligned} \tag{5.36}$$

which implies through the specific nondifferentiable potential:

$$Q = \frac{\lambda^2}{2}(d\tau)^{(4/D_F)-2}\frac{\partial^\mu \partial_\mu \sqrt{\rho}}{\sqrt{\rho}} = \left(U^\mu U_\mu - \lambda (d\tau)^{(2/D_F)-1}\partial^\mu U_\mu\right) \tag{5.37}$$

the specific nondifferentiable force:

$$F_\alpha = \frac{\lambda^2}{2}(d\tau)^{\left(\frac{4}{D_F}\right)-2}\partial_\alpha\left(\frac{\partial^\mu \partial_\mu \sqrt{\rho}}{\sqrt{\rho}}\right) = \left[U_\mu - \frac{\lambda}{2}(d\tau)^{\left(\frac{2}{D_F}\right)-1}\partial_\mu\right]\partial^\mu U_\alpha. \tag{5.38}$$

Thus, the first Eq. (7.26) takes the form:

$$\frac{\hat{d}\, V_\alpha}{d\tau} = \partial_\tau V_\alpha + V^\mu \partial_\mu V_\alpha = \frac{\lambda^2}{2}(d\tau)^{\left(\frac{4}{D_F}\right)-2}\partial_\alpha\left(\frac{\partial^\mu \partial_\mu \sqrt{\rho}}{\sqrt{\rho}}\right). \tag{5.39}$$

Since

$$V_\alpha = \lambda (d\tau)^{(2/D_F)-1}\partial_\alpha S, \tag{5.40}$$

which implies

$$V^\nu \partial_\nu V_\alpha = V^\nu \partial_\alpha V_\nu \tag{5.41}$$

and the relation (5.39) between the two becomes

$$\frac{\hat{d}\, V_\alpha}{d\tau} = \partial_\tau V_\alpha + V^\nu \partial_\alpha V_\nu - \frac{\lambda^2}{2}(d\tau)^{\left(\frac{4}{D_F}\right)-2}\partial_\alpha\left(\frac{\partial^\nu \partial_\nu \sqrt{\rho}}{\sqrt{\rho}}\right) = 0, \tag{5.42}$$

and more, for $\partial_\tau V_\alpha = 0$:

$$\partial_\alpha\left[V^\nu V_\nu - \lambda^2 (d\tau)^{(4/D_F)-2}\frac{\partial^\mu \partial_\mu \sqrt{\rho}}{\sqrt{\rho}}\right] = 0. \tag{5.43}$$

Now, by a suitable choice of constant integration and knowing that [1, 2]:

$$V^\mu V_\mu = \left(E^2 - \overleftarrow{\omega}^2 c^2\right)/m_0^2 \overleftarrow{\omega}^2.$$

We obtain the fractal energy expression in the form:

$$E = \pm\left[\left(m_0^2 \overleftarrow{\omega}^4 + p^2 \overleftarrow{\omega}^2 + \left(m_0 \overleftarrow{\omega}_\lambda\right)^2 (d\tau)^{(4/D_F)-2}\frac{\Box\sqrt{\rho}}{\sqrt{\rho}}\right]^{1/2}, \tag{5.44}$$

where

$$\Box = \partial^2/\partial x^2 + \partial^2/\partial y^2 + \partial^2/\partial z^2 - \overleftarrow{\omega}^{-2}\partial^2/\partial t^2.$$

For motions on the Peano curves, $D_F = 2$ at scale $\overleftarrow{\Lambda}_0 = \lambda/\overleftarrow{\omega} \equiv \hbar/m_0 c$, $\lambda = \hbar/m_0$, $\overleftarrow{\omega} \equiv c$ the fractal energy (7.44) is reduced to de Broglie's relation:

$$E = \pm\left[\left(m_0 c^2\right)^2 + p^2 c^2 + \frac{\Box\sqrt{\rho}}{\sqrt{\rho}}\right]^2 \tag{5.45}$$

In (5.49), $\hbar$ is the reduced Planck's constant, m_0 is the rest mass of the particle and c is the speed of light in the vacuum.

In such a context, the main conclusions are as follows:

(i) Assuming that in a Minkowski-type space–time, the motions of structural units take place on continuous but nondifferentiable curves, a scale relativity theory with an arbitrary constant fractal dimension is built;

(ii) Nondifferentiable geodesics on a space–time manifold and its diverse variants (the hydrodynamics one and the wave function) are obtained. Particularly, if the wave function is independent of the motion curve affine parameter and if we consider a scale resolution dependence on a fundamental length, then the nondifferentiable geodesics imply a nondifferentiable Klein–Gordon type equation. In this latter situation, the standard result (Klein–Gordon equation) is obtained for motions on the Peano curves at the Compton-type scale;

(iii) The fractal energy expression for motions on Peano curves at a Compton-type scale is reduced to the standard Relationships of de Broglie.

We note that every scale resolution is characterized by a fundamental length and by a specific velocity. From such a perspective, a unitary theory of interactions can be built.

References

1. Nottale, L. (1993). *Fractal space-time and microphysics: Towards a theory of scale relativity*. World Scientific.
2. Nottale, L. (2011). *Scale relativity and fractal space-time—A new approach to unifying relativity and quantum mechanics*. Imperial College Press.
3. Nottale, L., Scale relativity: A fractal matrix for organization in nature, EJTP 4, No. 16 (III), 15–102.
4. Mandelbrot, B. (1983). The fractal geometry of nature (Updated and Augm. Ed.), W.H. Freeman, New York.
5. Timofte, A., Casian-Botez, I., Scurtu, D., & Agop, M. (2011). System dynamics control through the fractal potential. *Acta Physica Polonica A, 119*(3), 304–311.
6. Niculescu, O., Nica, P., Gurlui, S., Forna, N., Casian Botez, I., Ioniţă, I., Constantin, B., & Bădărău, Gh. (2009). Experimental investigations of polymer plasma laser ablation. *Materiale Plastice, 46*(3), 336–338.
7. Nica, P. E., Vizureanu, P., Agop, M., Gurlui, S., Ziskind, M., Focşa, C., Forna, N., Ioannou, P. D., & Borsos, Z. (2009). Experimental and theoretical aspects of aluminium expanding laser plasma. *Japanese Journal of Applied Physics, 48*(6), 1–7.
8. Agop, M., Păun, V., Harabagiu, A., & El,. (2008). Naschie's theory and effects of nanoparticle clustering on the heat transport in nanofluids. *Chaos Solitons & Fractals, 37*(5), 1269–1278.

9. Agop, M., Forna, N., & Casian-Botez, I. (2008). New theoretical approach of the physical processes in nanostructures. *Journal of Computation and Theoretical Nanoscience, 5*(4), 483–489.
10. Gurlui, S., Agop, M., Strat, M., Băcăiţă, S., & Cerepaniuc, A. (2006). Some experimental and theoretical results on the on the anodic patterns in plasma discharge. *Physics of Plasmas, 13*(6), article id. 063503, 10.
11. Colotin, M., Pompilian, G. O., Nica, P., Gurlui, S., Păun, V., & Agop, M. (2009). Fractal transport phenomena through the scale relativity model. *Acta Physica Polonica A, 116*(2), 157–164.
12. Agop, M., Gavrilut, A., Crumpei, G., & Doroftei, B. (2014). Informational non-differentiable entropy and uncertainty relations in complex systems. *Entropy, 16*(11), 6042–6058. https://doi.org/10.3390/e16116042
13. Flake, G. W. (1998). *The computational beauty of nature*. MIT Press.
14. Mitchell, M. (2009). *Complexity: A guided tour*. Oxford University Press.
15. Winfree A. T. (2000). *The geometry of biological time*. Springer 2nd Edition, New York.
16. Luis, G. (1993). *Complex fluid*, Vol. 415. Springer.
17. Michel, O. D., & Thomas, B. G. (2012). *Mathematical modelling for complex fluids and flows*. Springer.
18. Hou, T. Y. (2009). *Multi-scale phenomena in complex fluids: Modelling*. World Scientific Publishing Company.
19. Cresson, J. (2003). Scale calculus and the Schrödinger equation. *Journal of Mathematical Physics, 44*(11), 4907–4938.
20. Phillips, A. C. (2003). *Introduction to quantum mechanics*. Wiley.
21. Agop, M., Gavriluţ, A., Ştefan, G., & Doroftei, B., Implications of non-differentiable entropy on a space-time manifold. *Entropy, 17*, 2184–2197; 10.3390\e17042184
22. de Broglie, L. (1966). *Thérmodynamique de la Particule Isolée ou Thérmodynamique Cachée de la Particul*. Gauthier - Villars.

Chapter 6
Some Applications of Spatial Multifractal Manifolds

In the present chapter, some applications in the description of various motions of scale relativity theory in the form of the Madelung-type scenario are given. From this perspective, the spatial coordinates are fractal/multifractal, while the temporal coordinates are not.

6.1 Kepler-Type Problems in the Madelung-Type Scenario

We consider a test particle in the scalar field $U = -\Theta/r$. The collective chaotic effect on the considered particle of all the other particles has a resulting motion on fractal curves. For example, a Levy-type motion with a "diffusion" coefficient λ may be chosen. Then, the first Eq. (4.40) gives

$$\partial^i \partial_i \sqrt{\rho} = -\frac{1}{2m_0\lambda^2 (dt)^{(4/D_F)-2}}\left(E - \frac{1}{2}m_0 V_D^i V_{Di} + \frac{\Theta}{r}\right)\sqrt{\rho},$$
$$\partial_i\left(\rho V_D^i\right) = 0, 0 < r < \infty, 0 \le \delta \le \pi, 0 \le \varphi \le 2\pi,$$
$$\rho(r = \infty, \delta, \varphi) = 0, \rho(r, \delta, \varphi)_{\delta \le \pi/2} = \rho(r, \delta + \pi/2, \varphi), \tag{6.1}$$

where E is the constant energy, m_0 is the rest mass of the test particle and Θ is the constant imposed by the interaction type (for details in the case of the fractal atom, see [1]).

We note that if the spatial manifold is operating simultaneously with multiple fractal dimensions, the fractal dimensions D_F can be substituted with a singularity spectrum $f(\alpha)$ of order α with $\alpha = \alpha(D_F)$.

The probability currents can flow in closed lines. Because of the azimuthal symmetry of complex systems, the real part of the complex speed potential has the following form:

M. Agop and S. A. Irimiciuc, *Multifractal Theory of Motion*,
https://doi.org/10.1007/978-981-97-6410-5_6

$$S = m\phi. \tag{6.2}$$

The first Eq. (4.32) corresponds to the real part of the complex speed field independent of ϕ:

$$\boldsymbol{V}_D = 2\lambda(dt)^{\left(\frac{2}{D_F}\right)-1}\frac{\boldsymbol{e}_\phi}{\mathrm{r}\sin\delta}m, m = 0, \pm1, \pm2, \ldots, \tag{6.3}$$

where $\boldsymbol{e}_\phi$ is the unit vector associated with the angle ϕ.

This statement is in agreement with the following conditions:

$$\begin{aligned}\oint m_0 V_D dr &= 2m_0\lambda(dt)^{(2/D_F)-1}m\oint\frac{e_\phi dr}{r\sin\delta}\\ &= 2m_0\lambda(dt)^{(2/D_F)-1}m\int_0^{2\pi}d\phi = 4\pi m_0\lambda(dt)^{(2/D_F)-1}m.\end{aligned} \tag{6.4}$$

By substituting Eqs. (6.3), (6.1) is reduced to the following equation:

$$\begin{aligned}\partial^i\partial_i\sqrt{\rho} = &-\frac{1}{2m_0\lambda^2(dt)^{(4/D_F)-2}}\left[E - 2m_0\lambda^2(dt)^{(4/D_F)-2}\right.\\ &\left.\times\left(\frac{m^2}{r^2\sin^2\delta}+\frac{\Theta}{r}\right)\right]\sqrt{\rho}.\end{aligned} \tag{6.5}$$

The variable ρ is independent of ϕ, i.e., $\rho = \rho(r, \delta)$, since $\boldsymbol{V}_D \| \boldsymbol{e}_\phi$ and $\nabla.(\rho\boldsymbol{V}_D) = 0 = \boldsymbol{V}_D \cdot \nabla\rho = 0$. By means of the statement:

$$\sqrt{\rho} = F(r)\Omega(\delta). \tag{6.6}$$

Equation (6.5) can be separated into differential equations with respect to r and δ:

$$\frac{r^2}{F}\left[\frac{1}{r^2}\frac{d}{dr}\left(r^2\frac{dF}{dr}\right)+\frac{1}{2m_0\lambda^2(dt)^{(4/D_F)-2}}\left(E+\frac{\Theta}{r}\right)F\right] = \mu \tag{6.7}$$

and:

$$\mu = -\frac{1}{\Omega}\left[\frac{1}{\sin\delta}\frac{d}{d\delta}\left(\sin\delta\frac{d\Omega}{d\delta}\right)-\frac{m^2}{\sin^2\delta}\Omega\right], \tag{6.8}$$

where μ is a separation parameter.

Under such conditions, the solutions of Eqs. (6.7) and (6.8) exist if the constants E and μ have the following eigenvalues:

$$\mu_l = l(l+1), l = 0, 1, 2 \tag{6.9}$$

and

$$E_n = -2m_0\lambda^2(dt)^{(4/D_F)-2}\frac{1}{a^2n^2}, n = 1, 2, 3 \tag{6.10}$$

with

$$a = m_0\frac{4\lambda^2(dt)^{(4/D_F)-2}}{\Theta}. \tag{6.11}$$

Thus, after normalization, one finds the following solution:

$$\begin{aligned} \boldsymbol{V}_{D(nl\ m)} &= 2\lambda(dt)^{(2/D_F)-1}\frac{\boldsymbol{e}_\phi}{r\sin\delta}m \\ \rho_{(nl\ m)} &= \left(\frac{2}{na}\right)^3\frac{(n-l-1)!}{2n[(n+l)!]^3}\frac{2l+1}{4\pi}\frac{(l-|m|)!}{(l+|m|)!}\left(\frac{2r}{na}\right)^{2l} \\ &\cdot\exp\left(-\frac{2r}{na}\right)\left[L_{n+l}^{2l+1}\left(\frac{2r}{na}\right)P_l^m(\cos\delta)\right]^2. \end{aligned} \tag{6.12}$$

In the previous relations, L_{n+l}^{2l+1} are the associated Laguerre polynomials, P_l^m are the associated Legendre polynomials [2], n is the equivalent of the principal quantum number, l is the equivalent of the orbital quantum number and m is the equivalent of the magnetic quantum number. For physical reasons, only the following combination of quantum numbers is acceptable:

$$0 \leq l \leq n-1, -l \leq m \leq +l. \tag{6.13}$$

In this context, the average distance is given by the following relation [3]:

$$< r >_{nl} = a\left[\frac{3}{2}n^2 - \frac{l}{2}(l+1)\right].$$

For the maximal value of the angular momentum, $l = n - 1$, the mean distance of the test particle becomes

$$< r >_n = a\left(n^2 + \frac{n}{2}\right)$$

The probability peak lies at the following values:

$$r_n(\text{ peak }) = an^2.$$

With the energy quantized as given in Eq. (6.10), the probability peaks for the semimajor axes lie at the following values:

$$a_n = an^2.$$

From Eq. (6.10) in the form:

$$\frac{E_n}{m_0} = -2\lambda^2 (dt)^{(4/D_F)-2} \frac{1}{a^2 n^2} = -\frac{1}{2}\left(\frac{w}{n}\right),^2$$

which results in the following velocity quantized law:

$$V_n = \frac{w}{n}, w = \frac{2\lambda(dt)^{(2/D_F)-1}}{a}.$$

The specific angular momentum is quantized:

$$\left(\frac{L}{m_0}\right) = L_0 l(l+1), L_0 = 4\lambda(dt)^{(2/D_F)-1}$$

and its projected component:

$$\frac{L_z}{m_0} = L_{0z} m, L_{0z} = 2\lambda(dt)^{(2/D_F)-1}, -l \leq m \leq l.$$

Since the Runge-Lenz vector $\boldsymbol{A}$:

$$\boldsymbol{A} = \boldsymbol{p} \times \boldsymbol{L} - \frac{\boldsymbol{r}}{r},$$

where $\boldsymbol{p}$ is the momentum vector, $\boldsymbol{L}$ is the angular momentum vector, and $\boldsymbol{r}$ is the position vector, which is a constant of motion in Kepler-type problems. Its projected component A_z is identified with the eccentricity e:

$$A_z \equiv e$$

and it is also quantized [2]:

$$A_z \equiv e = \frac{k}{n}, k = 0, 1, \ldots, n-1.$$

In Figs. 6.1, 6.2, 6.3, 6.4 and 6.5 the complete solutions of the fractal hydrodynamics model of the Kepler-type problems for various equivalent quantum numbers are given (three- and two-dimensional evolution plots).

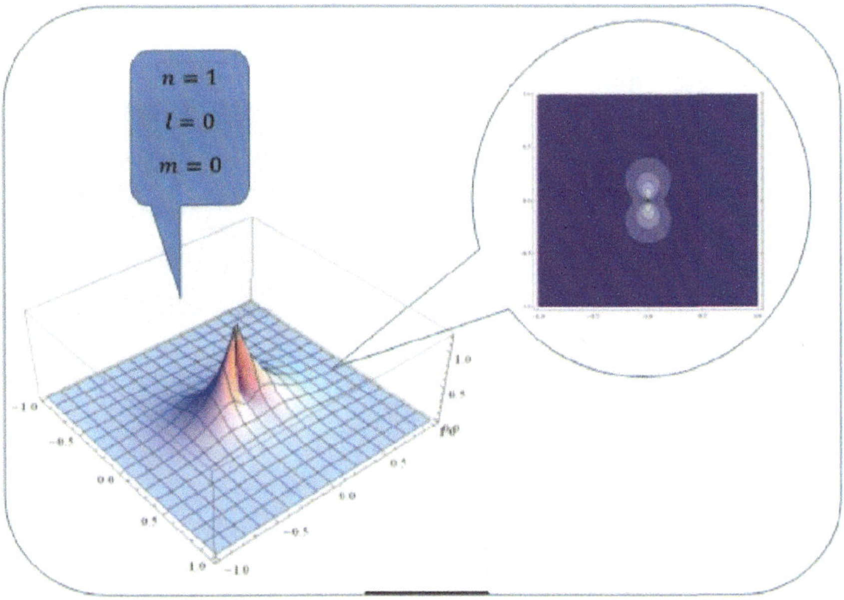

Fig. 6.1 Three- and two-dimensional (contour) curves dependencies of the state density in the one-body problem on the normalized coordinates $r/a = \left(x^2 + y^2\right)^{1/2}/a$, $\cos\delta = x/\left(x^2 + y^2\right)^{1/2}$ and equivalent quantum numbers $n = 1, l = 0, m = 0$

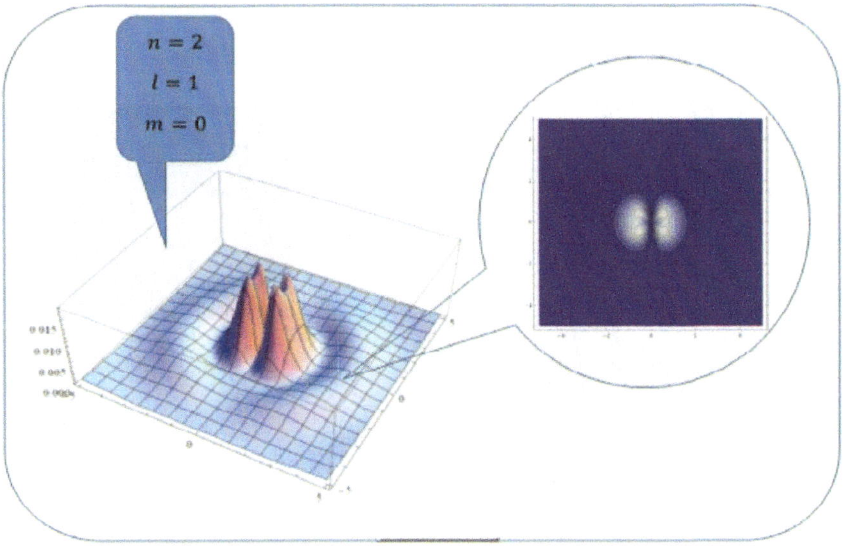

Fig. 6.2 Three- and two-dimensional (contour curves) Dependencies of the state's density in the one-body problem on normalized coordinates $r/a = \left(x^2 + y^2\right)^{1/2}/a$, $\cos\delta = x/\left(x^2 + y^2\right)^{1/2}$ and equivalent quantum numbers $n = 2, l = 1, m = 0$

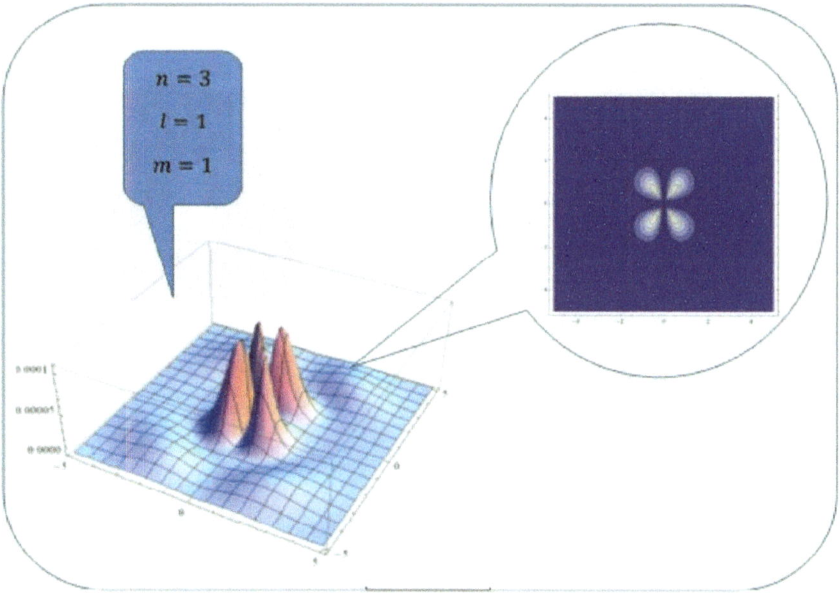

Fig. 6.3 Three- and two-dimensional (contour curves) dependencies of the state density in the one-body problem on the normalized coordinates $r/a = \left(x^2 + y^2\right)^{1/2}/a$, $\cos\delta = x/\left(x^2 + y^2\right)^{1/2}$ and equivalent quantum numbers $n = 3, l = 1, m = 1$

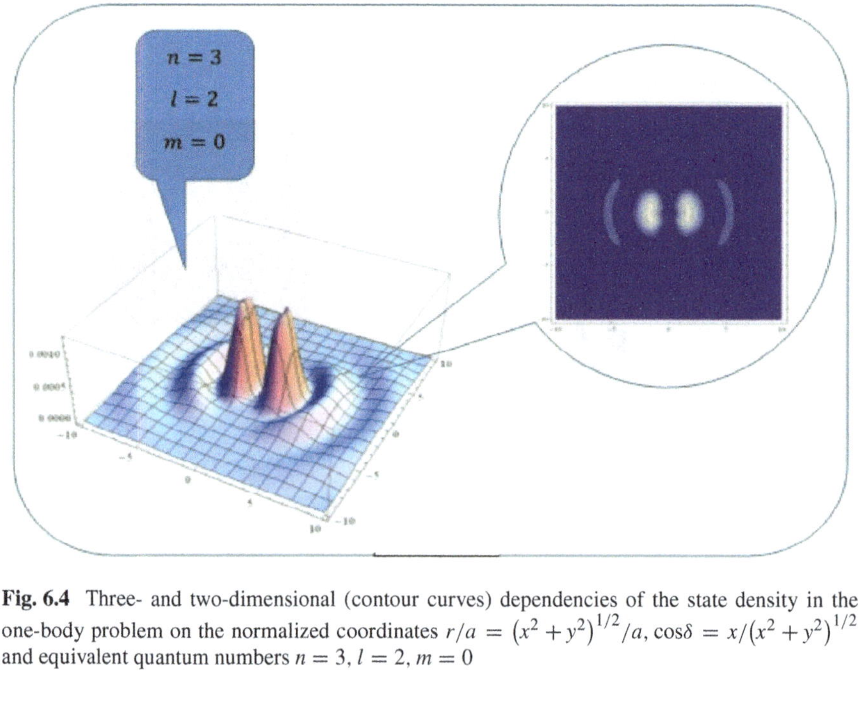

Fig. 6.4 Three- and two-dimensional (contour curves) dependencies of the state density in the one-body problem on the normalized coordinates $r/a = \left(x^2 + y^2\right)^{1/2}/a$, $\cos\delta = x/\left(x^2 + y^2\right)^{1/2}$ and equivalent quantum numbers $n = 3, l = 2, m = 0$

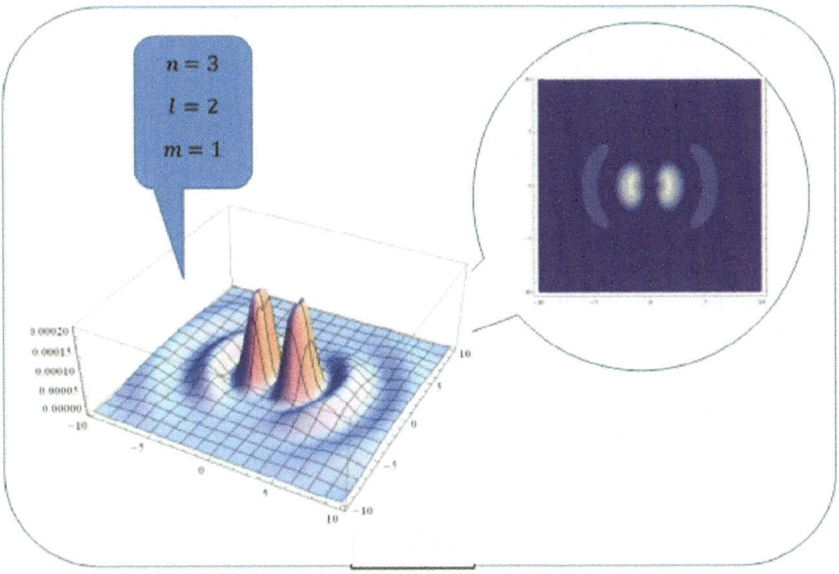

Fig. 6.5 Three- and two-dimensional (contour curves) dependencies of the state density in the one-body problem on the normalized coordinates $r/a = \left(x^2 + y^2\right)^{1/2}/a$, $\cos\delta = x/\left(x^2 + y^2\right)^{1/2}$ and equivalent quantum numbers $n = 3, l = 2, m = 1$

6.2 Harmonic Oscillator-Type Problems in the Madelung-Type Scenario

The one-dimensional harmonic oscillator in a fractal space–time can be assimilated to a complex system's structural unit in a spatially symmetric potential energy field of the form:

$$U(x) = \frac{1}{2}m_0\Omega^2x^2, \; -\infty \leq x \leq +\infty. \tag{6.14}$$

In the stationary state, Eq. (4.34) yields

$$\rho(x)V_D(x) = c = \text{const.} \tag{6.15}$$

The same equation is used for the boundary conditions:

$$\rho(-\infty) = 0, \rho(+\infty) = 0 \tag{6.16}$$

implies $c = 0$. Hence

$$V_D(x) = 0. \tag{6.17}$$

Accordingly, Eq. (4.40) is applied, which leads (under consideration of Eq. (6.14)) to the linear boundary-value problem:

$$\begin{aligned} &\frac{d^2\sqrt{\rho}}{dx^2} + \left[\left(\frac{E}{2m_0D^2}\right) - \left(\frac{\Omega}{2D}\right)^2 x^2\right]\sqrt{\rho} = 0, \\ &-\infty \le x \le +\infty, \rho(x = -\infty) = \rho(x = +\infty) = 0 \\ &D = \lambda(dt)^{(2/D_F)-1}. \end{aligned} \tag{6.18}$$

By introducing the nondimensional variable:

$$\xi = \sqrt{\frac{\Omega}{2D}}x, \tag{6.19}$$

Equation (6.18) takes the form:

$$\begin{aligned} &\frac{d^2\sqrt{\rho}}{d\xi^2} + \left(\frac{E}{m_0D\Omega} - \xi^2\right)\sqrt{\rho} = 0, \\ &-\infty \le \xi \le +\infty, \rho(\xi = -\infty) = \rho(\xi = +\infty) = 0. \end{aligned} \tag{6.20}$$

A solution $\rho(\xi) = \rho_n(\xi)$ can be found if the total energy of the complex system assumes certain values, $E = E_n$. Using the method from [4], one finds:

$$\rho_n(\xi) = \sqrt{\frac{\Omega}{2\pi D}\frac{1}{2^n n!}}H_n^2(\xi)e^{-\xi^2} \tag{6.21}$$

with

$$E_n = 2m_0D\Omega\left(n + \frac{1}{2}\right), \tag{6.22}$$

where $H_n(\xi)$ are Hermite polynomials [2].

In particular, for $D_F = 2$ and $D = \hbar/2m_0$, we refined the results of quantum mechanics [2]:

$$\begin{aligned} \Psi_n(x) &= \left(\tfrac{m_0\Omega}{\pi\hbar}\right)^{1/4}\frac{1}{\sqrt{2^n n!}}e^{-\frac{m_0\Omega}{2\hbar}x^2}H_n\left(x\sqrt{\tfrac{m_0\Omega}{\hbar}}\right), \\ E_n &= \left(n + \tfrac{1}{2}\right)\hbar\Omega, n = 0, 1, 2, \ldots, \end{aligned}$$

where Ψ_n are the normalized wave functions of the stationary states and E_n are the discrete eigenvalues associated with those states.

Therefore, Eqs. (6.17) and (6.21) specify the following:

(i) At the differentiable scale, the differentiable part of the complex speed field of the complex system, V_D^l, is null—Eq. (6.17). In agreement with the second

Eq. (4.32), $S = \text{const.}$ (the complex system is coherent—the structural units of the complex system have the same phase);

(ii) At the nondifferentiable scale, the imaginary (nondifferentiable or fractal) part of the complex speed field of the complex system V_F^l is nonzero and can be expressed as follows [5]:

$$V_F(x) = D\frac{d\ln\rho}{dx} = D\frac{d\ln\rho}{d\xi}\cdot\frac{d\xi}{dx} = \sqrt{2D\Omega}\left[\frac{H_n\prime(\xi)}{H_n(\xi)} - \xi\right] \\ = \sqrt{2D\Omega}\left[2n\frac{H_{n-1}(\xi)}{H_n(\xi)} - \xi\right]. \tag{6.23}$$

where we use the following relationship [2]:

$$H_n\prime = \frac{dH_n(\xi)}{d\xi} = 2nH_{n-1}(\xi). \tag{6.24}$$

Then, the fractal one-dimensional potential [4] is

$$Q_n = -\frac{m_0V_F^2}{2} - m_0D\frac{dV_F}{dx}$$

with

$$-\frac{m_0V_F^2}{2} = -m_0D\Omega\left[4n^2\frac{H_{n-1}^2(\xi)}{H_n^2(\xi)} + \xi^2 - 4n\xi\frac{H_{n-1}(\xi)}{H_n(\xi)}\right]$$

and

$$-m_0D\frac{dV_F}{dx} = -m_0D\Omega\left[2n\frac{H_{n-1}\prime(\xi)H_n(\xi) - H_n\prime(\xi)H_{n-1}(\xi)}{H_n^2(\xi)} - 1\right] \\ = m_0D\Omega\left[4n(n-1)\frac{H_{n-2}(\xi)}{H_n(\xi)} - 4n^2\frac{H_{n-1}^2(\xi)}{H_n^2(\xi)} - 1\right]$$

involving the explicit form:

$$Q_n = -m_0D\Omega\left[4n^2\frac{H_{n-1}^2(\xi)}{H_n^2(\xi)} + \xi^2 - 4n\xi\frac{H_{n-1}(\xi)}{H_n(\xi)}\right. \\ \left. +4n(n-1)\frac{H_{n-2}(\xi)}{H_n(\xi)} - 4n^2\frac{H_{n-1}^2(\xi)}{H_n^2(\xi)} - 1\right] \\ = -m_0D\Omega\left[\xi^2 - 1 + 2n\frac{-2\xi H_{n-1}(\xi) + 2(n-1)H_{n-2}(\xi)}{H_n(\xi)}\right] \\ = m_0D\Omega(2n + 1 - \xi^2), \tag{6.25}$$

where we use the following relationship [2]:

$$H_n(\xi) - 2\xi H_{n-1}(\xi) + 2(n-1)H_{n-2}(\xi) \equiv 0. \tag{6.26}$$

Finally, the results are

$$E_n = U + Q_n = m_0 D\Omega\xi^2 + m_0 D\Omega\left(2n + 1 - \xi^2\right) \equiv 2m_0 D\Omega\left(n + \frac{1}{2}\right). \tag{6.27}$$

Therefore, the harmonic oscillator in a fractal space is equivalent to a coherent fractal fluid whose particles move on stationary trajectories that satisfy condition (6.27).

The identity of relations (6.22) and (6.27) shows that the "observable" in the form of energy (generated both by the fractal component of the complex speed through the fractal potential and by the external potential) is quantified. The zero value of the real (differentiable) part of the complex speed field corresponds to a superconductor or a superfluid type behavior of the fractal fluid, and the nonzero value of the imaginary (fractal) part selects, through condition (6.27), the stationary "trajectories" of the fluid particles. Momentum transfer is achieved only through the fractal components of the complex speed field.

The previous result can be generalized for an n-dimensional harmonic oscillator. In this context, we show in Figs. 6.6, 6.7, 6.8, 6.9 and 6.10 the density of states of the two-dimensional harmonic oscillator for various quantum states induced by the quantum numbers n_ξ and n_η (the three-dimensional and contour plot dependences, respectively).

6.3 Free Particles in the Madelung-Type Scenario

We use Eqs. (4.33)–(4.35) to study the dynamics of free perturbation. Due to their nonlinearity, obtaining an analytical solution in the general case can be difficult. However, there are special situations in which this is possible. Indeed, let us consider the one-dimensional case of Eqs. (4.33)–(4.35) in the absence of any external constraint, $U = 0$, i.e., [6–8]

$$\partial_i V_D + V_D \partial_x V_D = 2\lambda^2 (dt)^{(4/D_F)-2} \rho^{-1/2} \partial_{xx} \rho^{1/2} \tag{6.28}$$

$$\partial_t \rho + \partial_x (\rho V_D) = 0 \tag{6.29}$$

with the initial conditions:

$$V_D(x, t = 0) = c \tag{6.30}$$

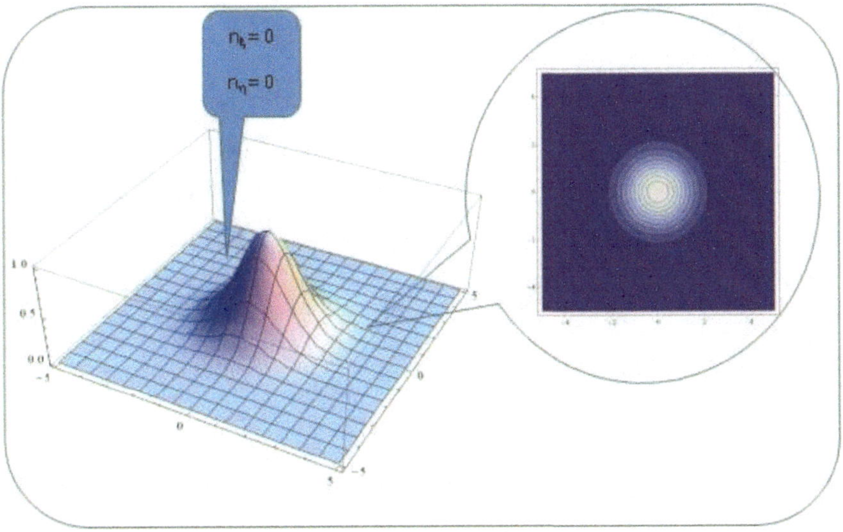

Fig. 6.6 The state densities of the two-dimensional harmonic oscillator (three-dimensional and contour plot dependencies) for quantum number $n_\xi = 0, n_\eta = 0$

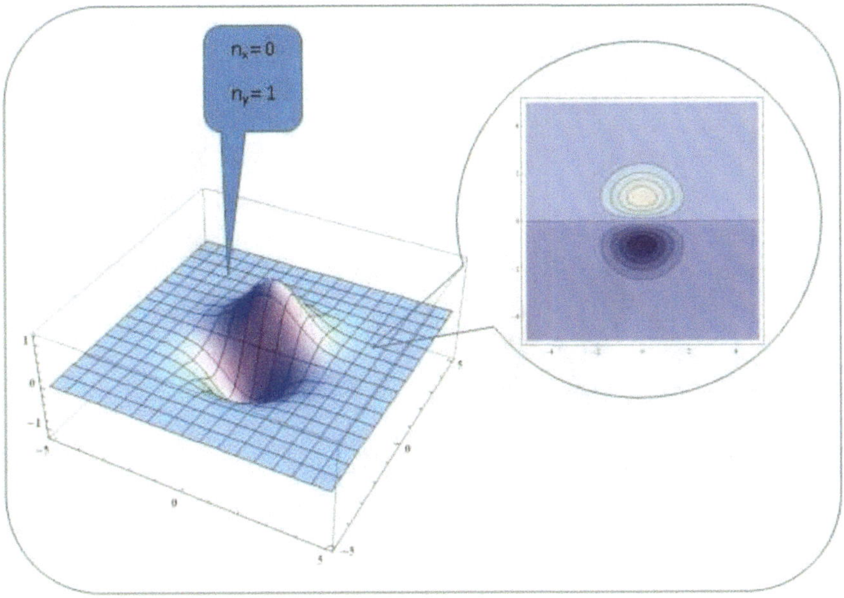

Fig. 6.7 The state densities of the two-dimensional harmonic oscillator (three-dimensional and contour plot dependencies) for quantum number $n_\xi = 0, n_\eta = 1$

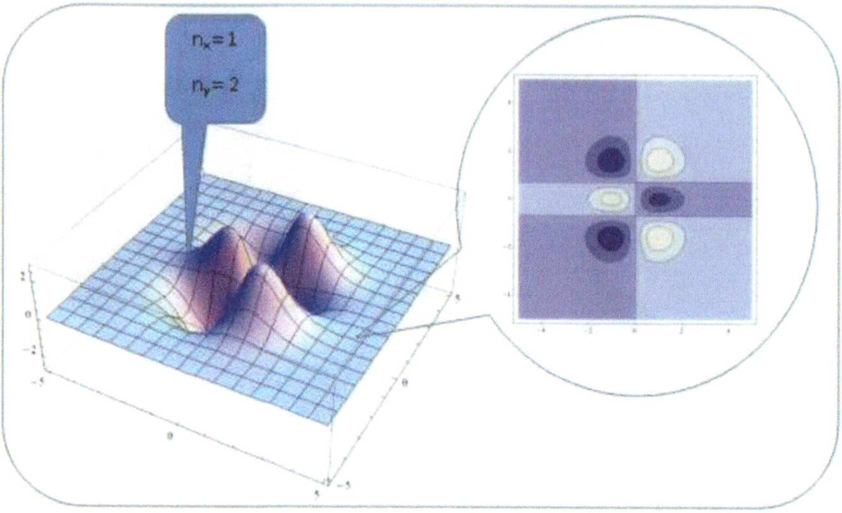

Fig. 6.8 The two-dimensional density of states harmonic oscillators (three-dimensional and contour plot dependencies) for quantum number $n_\xi = 1, n_\eta = 2$

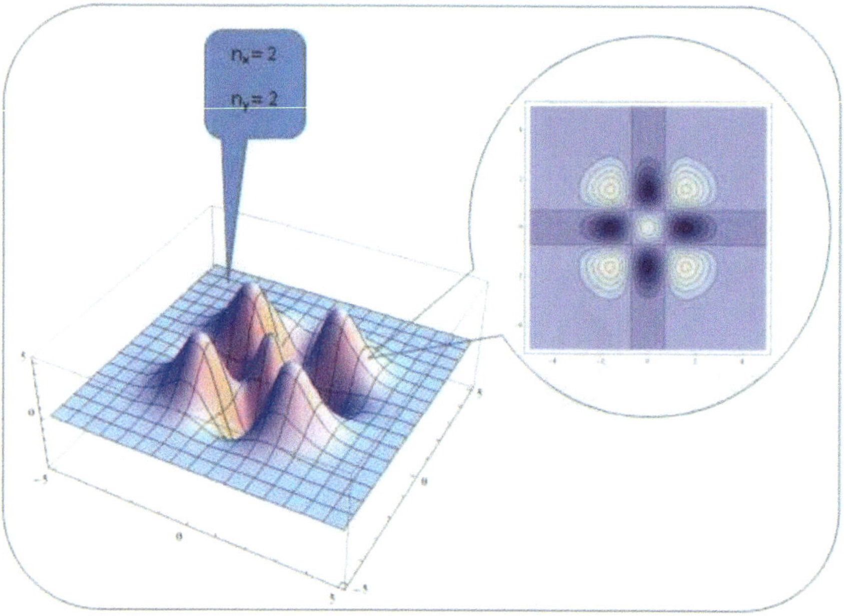

Fig. 6.9 The two-dimensional density of states harmonic oscillators (three-dimensional and contour plot dependencies) for quantum number $n_\xi = 2, n_\eta = 2$

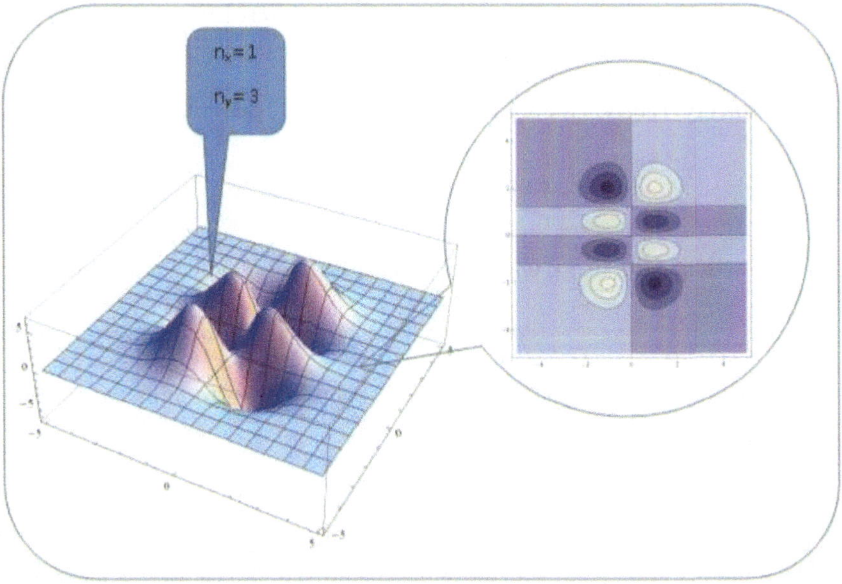

Fig. 6.10 The two-dimensional density of states harmonic oscillators (three-dimensional and contour plot dependencies) for quantum number $n_\xi = 1, n_\eta = 3$

$$\rho(x, t = 0) = \rho_0 \exp\left[-\left(\frac{x}{a}\right)^2\right] \tag{6.31}$$

and the boundary one

$$V_D(x = ct, t) = c \tag{6.32}$$

$$\rho(x = -\infty, t) = \rho(x = \infty, t) = 0. \tag{6.33}$$

For any $t > 0$ or $t < 0$, $< \partial_x Q >= 0$ results (for details, see [4]). Since $\langle x \rangle = ct$ in the absence of external forces, Eq. (6.28) can be separated into

$$\partial_x\left(\rho^{-1/2}\partial_{xx}\rho^{1/2}\right) = \frac{2}{a(t)^2}(x - ct) \tag{6.34}$$

and

$$\partial_t V_D + V_D \partial_x V_D = \left[\frac{2\lambda}{a(t)}\right]^2 (dt)^{(4/D_F)-2}(x - ct). \tag{6.35}$$

The integration of Eq. (6.34) gives, taking into consideration the boundary condition from (6.33), a quadratic solution in $(x - ct)$. i.e.:

$$\rho(x,t) = \frac{1}{[\pi a(t)]^{1/2}} \exp\left[-\frac{(x-ct)^2}{a(t)}\right]. \tag{6.36}$$

This function satisfies the initial condition (6.31) if the initial value of $a(t)$ is chosen as:

$$a(t=0) = \alpha^2. \tag{6.37}$$

The insertion of Eq. (6.36) into the continuity Eq. (6.29) indicates that for $x = ct$:

$$(2a)^{-1} d_t a = (\partial_x V_D)_{x=ct}. \tag{6.38}$$

Then, the differential equation for $a(t)$ is obtained by performing the operation $(\partial_x)_{x=ct}$ on Eq. (6.35):

$$a d_{tt} a - \frac{1}{2}(d_t a)^2 = 8\lambda^2 (dt)^{(4/D_F)-2}. \tag{6.39}$$

The solution of Eq. (6.39) with the initial condition (6.37) (satisfying the requirement that $\rho(x,t)$ be real, which is met by the symmetry condition $a(t) = a(-t)$, Eq. (6.36)) is

$$a(t) = \alpha^2 + \left[\frac{2\lambda (dt)^{(2/D_F)-1}}{\alpha}(t)\right]^2. \tag{6.40}$$

According to Eqs. (6.36) and (6.40), the state density is Gaussian in a form with a time-dependent distribution parameter $a(t)$ and spreads with the "classical" particle velocity c:

$$\rho(x,t) = \frac{\pi^{-1/2}}{\left\{\alpha^2 + \left[\frac{2\lambda (dt)^{(2/D_F)-1}}{\alpha} t\right]^2\right\}^{1/2}} \times \exp\left\{-\frac{(x-ct)^2}{\alpha^2 + \left[\frac{2\lambda (dt)^{(2/D_F)-1}}{\alpha} t\right]^2}\right\}. \tag{6.41}$$

Similarly, the integration of Eq. (6.28) with the initial condition (6.30) and the boundary condition (6.32) gives the velocity field of the particle at the differentiable scale:

$$V_D = \frac{c\alpha^2 + \left[\frac{2\lambda (dt)^{(2/D_F)-1}}{\alpha}\right]^2 tx}{\alpha^2 + \left[\frac{2\lambda (dt)^{(2/D_F)-1} t}{\alpha}\right]^2}. \tag{6.42}$$

Equations (6.41) and (6.42) represent the fractal hydrodynamic solutions of the free particles at the differentiable scale.

From this point on, we can define the following dynamic variables:

(i) Current density:

$$j(x,t) = \rho(x,t)V_D(x,t) = \frac{c\alpha^2 + \left(\frac{2\lambda(dt)^{(2/D_F)-1}}{\alpha}\right)^2 tx}{\left\{\alpha^2 + \left[\frac{2\lambda(dt)^{(2/D_F)-1}}{\alpha}t\right]^2\right\}^{3/2}} \times \exp\left\{-\frac{(x-ct)^2}{\alpha^2 + \left[\frac{2\lambda(dt)^{(2/D_F)-1}}{\alpha}t\right]^2}\right\}. \tag{6.43}$$

(ii) Specific fractal potential:

$$Q = -2\lambda^2(dt)^{(4/D_F)-2}\left\{\frac{(x-ct)}{\alpha^2 + \left[\frac{2\lambda(dt)^{(2/D_F)-1}}{\alpha}t\right]^2}\right\}^2; \tag{6.44}$$

(iii) Specific fractal force:

$$F(x,t) = -\partial_x Q = \lambda^2(dt)^{(4/D_F-2)}\frac{(x-ct)}{\left\{\alpha^2 + \left[\frac{2\lambda(dt)^{(2/D_F)-1}}{\alpha}t\right]^2\right\}^2}. \tag{6.45}$$

For $t > 0$ or $t < 0$, the dynamic variables defined in Eqs. (6.41)–(6.45) are inhomogeneous in x and t, while for $x = ct$, they become

$$V_D(x = ct, t) \to c, \tag{6.46}$$

$$\rho(x = ct, t) \to \frac{\pi^{-1/2}}{\left\{\alpha^2 + \left[\frac{2\lambda(dt)^{(2/D_F)-1}}{\alpha}t\right]^2\right\}^{1/2}} \tag{6.47}$$

$$j(x = ct, t) \to \frac{\pi^{-1/2}c}{\left\{\alpha^2 + \left[\frac{2\lambda(dt)^{(2/D_F)-1}}{\alpha}t\right]^2\right\}^2} \tag{6.48}$$

$$Q(x = ct, t) \to 0, \tag{6.49}$$

$$F(x = ct, t) \to 0. \tag{6.50}$$

The previous relations can be greatly simplified by introducing the following nondimensional coordinates:

$$\xi = \frac{x}{\alpha}, \tau = \frac{tc}{\alpha}, 2\lambda = \alpha c. \tag{6.51}$$

A new type of normalization is thus generated, dependent both on the exterior "constraint" through the α parameter, related to the half-width of the position Gaussian, and on certain characteristics induced by this constraint, such as the particle's velocity c with respect to the "observable" plane.

Therefore, the following normalized variables can be defined:

(i) Normalized velocity (see Fig. 6.3a–c):

$$V(\xi, \tau) = \frac{V(\xi, \tau)}{c} = \frac{1 + (dt)^{\left(\frac{4}{D_F}\right)-2}\xi\tau}{1 + (dt)^{\left(\frac{4}{D_F}\right)-2}\tau^2}. \tag{6.52}$$

(ii) Normalized state density (see Fig. 6.4a–c):

$$N(\xi, \tau) = \pi^{1/2}\alpha\rho(\xi, \tau) = \left[1 + (dt)^{(4/D_F)-2}\tau^2\right]^{-1/2} \times \\ \times \exp\left[-\frac{(\xi - \tau)^2}{1 + (dt)^{(4/D_F)-2}\tau^2}\right]. \tag{6.53}$$

(ii) Normalized current density (see Fig. 6.5a–c):

$$J(\xi, \tau) = \frac{\pi^{1/2}\alpha}{c}\rho(\xi, \tau)V_D(\xi, \tau) = \frac{1 + (dt)^{(4/D_F)-2}\xi\tau}{\left[1 + (dt)^{(4/D_F)-2}\tau^2\right]^{3/2}} \\ \times \exp\left[-\frac{(\xi - \tau)^2}{1 + (dt)^{(4/D_F)-2}\tau^2}\right]. \tag{6.54}$$

(iii) Normalized fractal potential:

$$q(\xi, \tau) = \frac{\alpha^2 Q(\xi, \tau)}{2\lambda^2} = (dt)^{(4/D_F)-2}\left[\frac{(\xi - \tau)}{1 + (dt)^{(4/D_F)-2}\tau^2}\right]^2. \tag{6.55}$$

(iv) Normalized fractal force:

$$f(\xi, \tau) = \frac{\alpha^3 F(\xi, \tau)}{2\lambda^2} = (dt)^{(4/D_F)-2}\frac{(\xi - \tau)}{\left[1 + (dt)^{(4/D_F)-2}\tau^2\right]}. \tag{6.56}$$

(v) Normalized characteristic length:

$$\Lambda(\xi,\tau)=\left[\frac{q(\xi,\tau)}{N(\xi,\tau)}\right]^{1/2}=(dt)^{(2/D_F)-1}\frac{(\xi-\tau)}{\left[1+(dt)^{(4/D_F)-2}\tau^2\right]^{3/2}}$$
$$\times\exp\left[-\frac{1}{2}\frac{(\xi-\tau)^2}{1+(dt)^{(4/D_F)-2}\tau^2}\right]. \tag{6.57}$$

In the particular case $\xi=\tau$, relations (6.52) and (6.57) are reduced to

$$V(\xi=\tau,\tau)\to 1, \tag{6.58}$$

$$N(\xi=\tau,\tau)\to\left[1+(dt)^{(4/D_F)-2}\right]^{-1/2}, \tag{6.59}$$

$$J(\xi=\tau,\tau)\to\left[1+(dt)^{(4/D_F)-2}\right]^{-1/2} \tag{6.60}$$

$$q(\xi=\tau,\tau)\to 0, \tag{6.61}$$

$$f(\xi=\tau,\tau)\to 0, \tag{6.62}$$

$$\Lambda(\xi=\tau,\tau)\to 0. \tag{6.63}$$

By accepting the "scale superposition principle" [3] in Figs. 6.11, 6.12, 6.13, 6.14, 6.15, 6.16, 6.17, 6.18, 6.19 and 6.20, three- and two-dimensional (contour curves) depend on the spatial variable ξ and temporal variable τ both for individual current density $J_1(\xi,\tau)$ at the local scale $\mu_1=$ const. $J_2(\xi,\tau)$ at the local scale $\mu_2=$ const. and for the total current density $J_t(\xi,\tau)$ at the global scale, $\mu_1+\mu_2=$ const. by means of the scale superposition principle are given. Moreover, in Fig. 6.21, the temporal dependencies both for the individual current density $J_1(\tau)$ at the local scale μ_1, $J_2(\tau)$ at the local scale μ_2 and for the total current density $J_t(\tau)$ at the global scale $\mu_1+\mu_2$ are given.

6.4 Uncertainty Relations in the Madelung-Type Scenario

Usually, the uncertainty relations are assimilated to certain functional constraints imposed by the complex system's internal structure. In the fractal hydrodynamic model [9–13], the uncertainty relations result almost naturally from the specific fractal momentum perturbations associated with a specific fractal potential. Indeed, using identity:

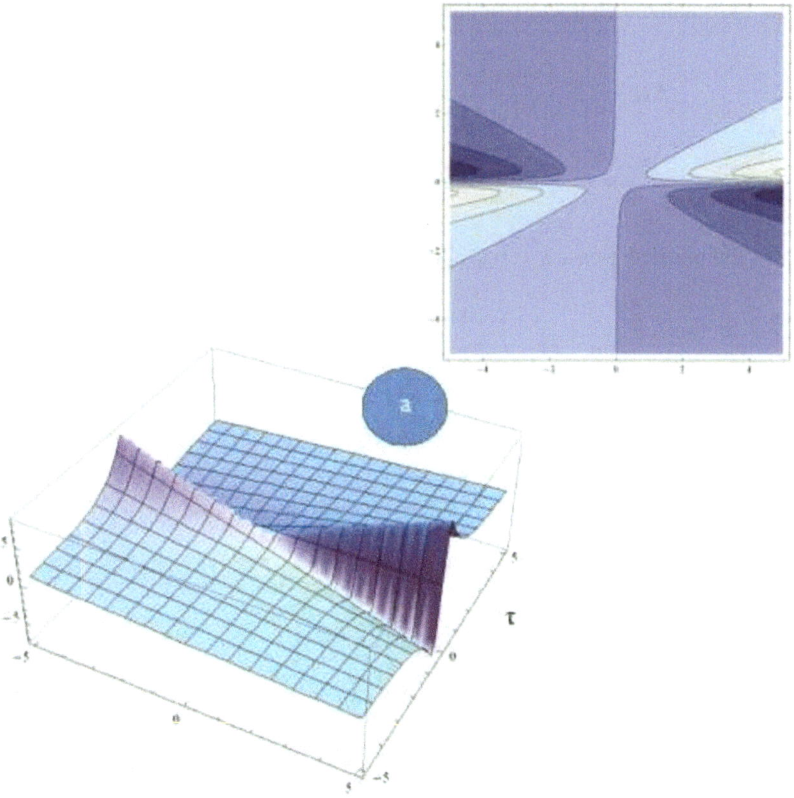

Fig. 6.11 Three- and two-dimensional (contour curves) the dependences $V(\xi, \tau, \mu = \text{const.})$

$$-\rho\nabla Q = \nabla\left[\rho\lambda^4(dt)^{(4/D_F)-2}\nabla\nabla\ln\rho\right] \tag{6.64}$$

the specific fractal momentum conservation law (4.33) gives

$$\nabla\cdot[\rho(\boldsymbol{V}_D\boldsymbol{V}_D)] = \nabla\cdot\left[\rho^2\lambda^4(dt)^{(4/D_F)-2}\nabla\nabla\ln\rho\right] + \ldots \tag{6.65}$$

Accordingly, the fractal stresses are potentially equivalent to specific fractal momentum stresses $p_i p_j$ imparted to the fractal hydrodynamic fluid associated with the structural unit:

$$\begin{aligned}\boldsymbol{pp} &= -\lambda^4(dt)^{(4/D_F)-2}\nabla\nabla\ln\rho \\ &= -\lambda^4(dt)^{(4/D_F)-2}\left(\frac{\nabla\nabla\rho}{\rho} - \frac{\nabla\rho}{\rho}\frac{\nabla\rho}{\rho}\right).\end{aligned} \tag{6.66}$$

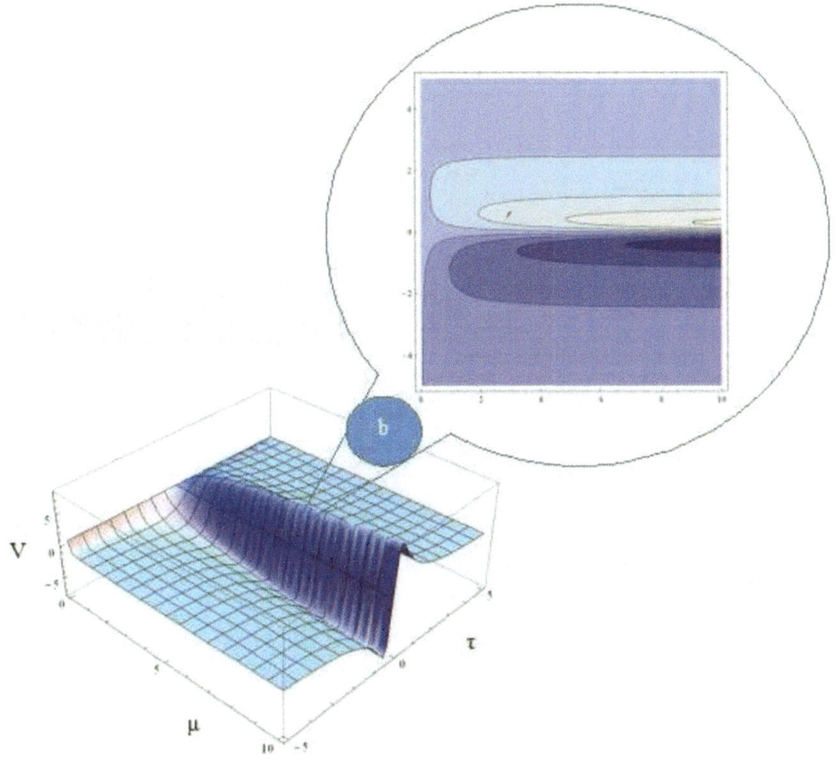

Fig. 6.12 Three- and two-dimensional (contour curves) dependences $V(\xi = \text{const.}, \tau, \mu)$

The expectation values of the specific fractal momentum stresses $p_i p_j$ represent the observable specific fractal momentum stresses of the structural unit. According to Eq. (6.65),

$$\langle \boldsymbol{pp} \rangle = \lambda^4 (dt)^{(4/D_F)-2} \iiint \rho \nabla \ln \rho \nabla \ln \rho d\boldsymbol{r} \tag{6.67}$$

since

$$\iiint \nabla\nabla\rho dr = \oint \nabla\rho df = \overleftrightarrow{0}. \tag{6.68}$$

The specific fractal momentum stresses $p_i p_j$, shown in Eq. (6.65), are due to unobservable (the first term) and observable (the second term) specific fractal stresses. The observable specific fractal momentum stresses are given by the dyad:

$$\boldsymbol{qq} = \lambda^4 (dt)^{\left(\frac{4}{D_F}\right)-2} \nabla \ln \rho \nabla \ln \rho, \boldsymbol{qq} \neq \overleftrightarrow{0}. \tag{6.69}$$

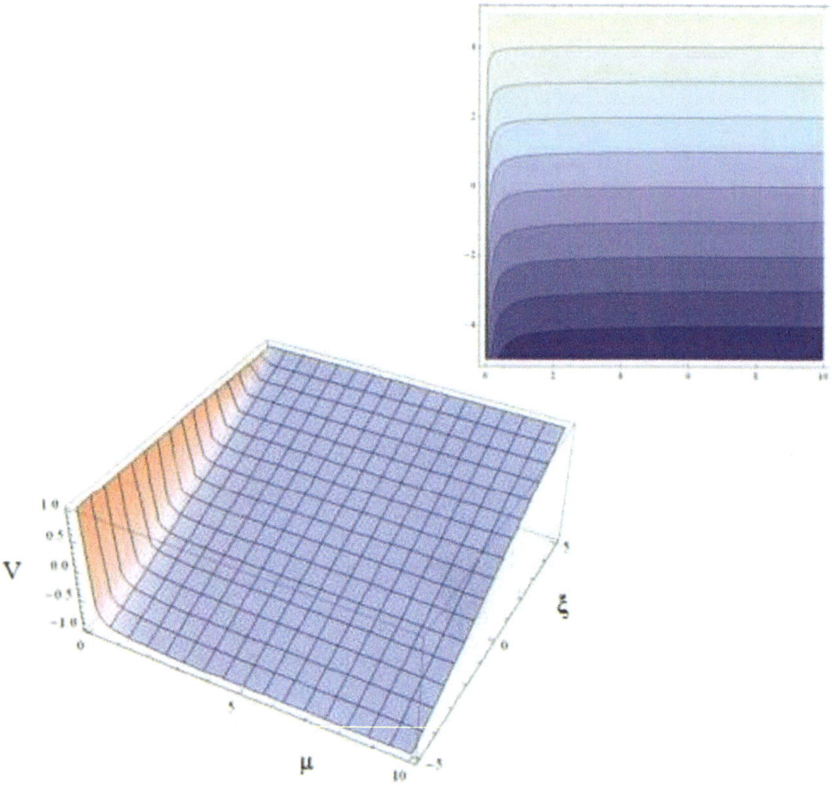

Fig. 6.13 Three- and two-dimensional (contour curves) the dependences $V(\xi, \tau = \text{const.}, \mu)$

They determine the observable fractal uncertainties Δx_{ij} of the conjugate components of the fractal position tensor $\boldsymbol{rr}$ of the structural unit. Thus, one finds from Eq. (6.69) the fractal relation:

$$<q_i q_j><\Delta x_{ij}>^2 = \lambda^4 (dt)^{(4/D_F)-2} \varepsilon_{ij}^2(s), \tag{6.70}$$

where

$$< q_i q_j >= \lambda^4 (dt)^{(4/D_F)-2} \iiint \rho(\partial_i \ln\rho \partial^i \ln\rho) d\boldsymbol{r},$$
$$< \Delta x_{ij} >^2 = < (x_i - < x_i >)(x_j - < x_j >) > .$$

$\varepsilon_{ij}(s)$ is a function of the set of numbers (the equivalent of quantum numbers at the fractal scale) specifying the fractal state of the complex system. For complex systems with a separable distribution fractal function $\rho(\boldsymbol{r}, t, dt) = \rho_1(x_1, t, dt)\rho_2(x_2, t, dt)\rho_3(x_3, t, dt)$, the nondiagonal variances vanish, $\Delta x_{ij} = 0$ for $i \neq j$. In this case,

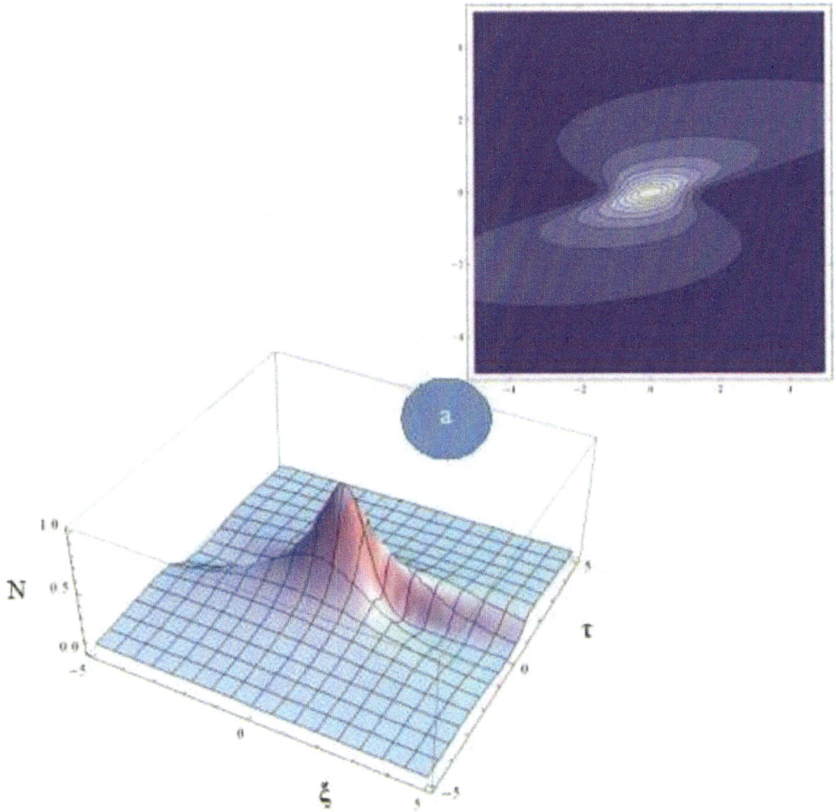

Fig. 6.14 Three- and two-dimensional (contour curves) the dependences $N(\xi, \tau, \mu = \text{const.})$

it results:

$$< q_i^2 >^{1/2} \Delta x_i = \lambda^2 (dt)^{(2/D_F)-1} \varepsilon_i(s), \tag{6.71}$$

where

$$< q_i^2 >= \lambda^4 (dt)^{(4/D_F)-2} \iiint \rho\left(\partial_i \ln\rho \, \partial^i \ln\rho\right) d\boldsymbol{r}, \tag{6.72}$$

$$< \Delta x_i >^2 = < (x_i - < x_i >)^2 > . \tag{6.73}$$

Equation (6.69) is the tensor formulation of the fractal hydrodynamic uncertainty relations.

The application of fractal hydrodynamic uncertainty relations to some complex concrete systems and the evaluation of the function of the fractal state are demonstrated in the following examples:

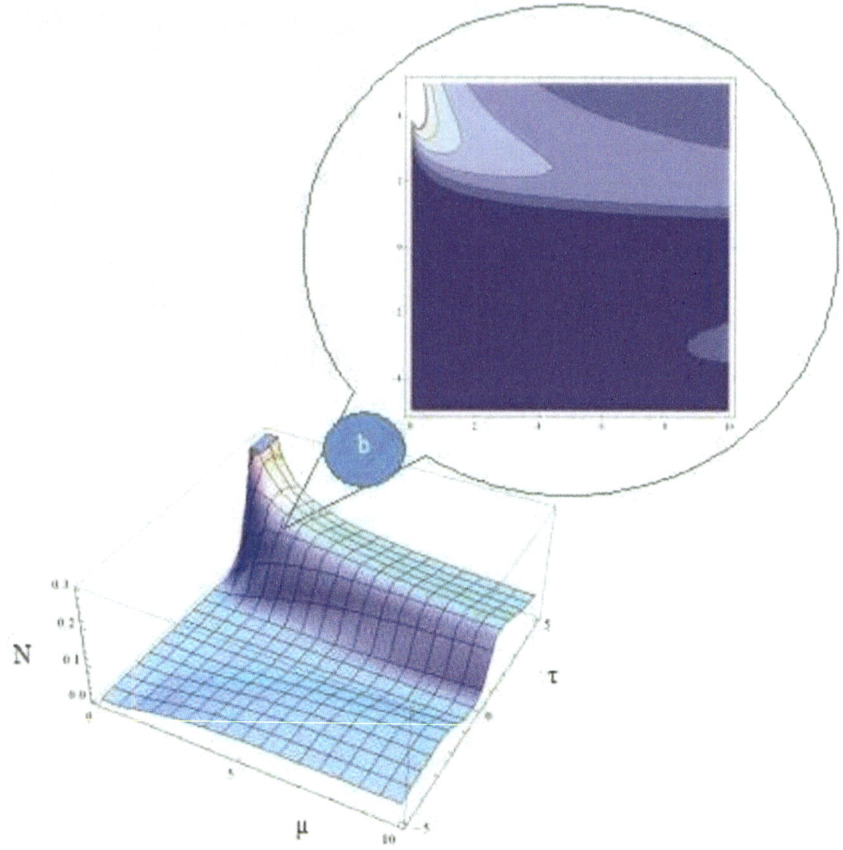

Fig. 6.15 Three- and two-dimensional (contour curves) dependences $N(\xi = \text{const.}, \tau, \mu)$

(i) By means of the solution for the Kepler-type problems, Eq. (6.12) is used to verify that

$$< q_r^2 >= \left[\frac{2m_0\lambda(dt)^{(2/D_F)-1}}{na}\right]^2\left(1 - \frac{2l}{n}\frac{l+1}{n+1}\right) \tag{6.74}$$

and

$$< (r- < r >)^2 >= \left(\frac{1}{2}a\right)^2\left[n^2(n^2+2) - l^2(l+1)^2\right]. \tag{6.75}$$

Accordingly, the uncertainty relation, Eq. (6.74), becomes, for the r components of the dynamical variables of the test particle in the potential $U = -\Theta/r$:

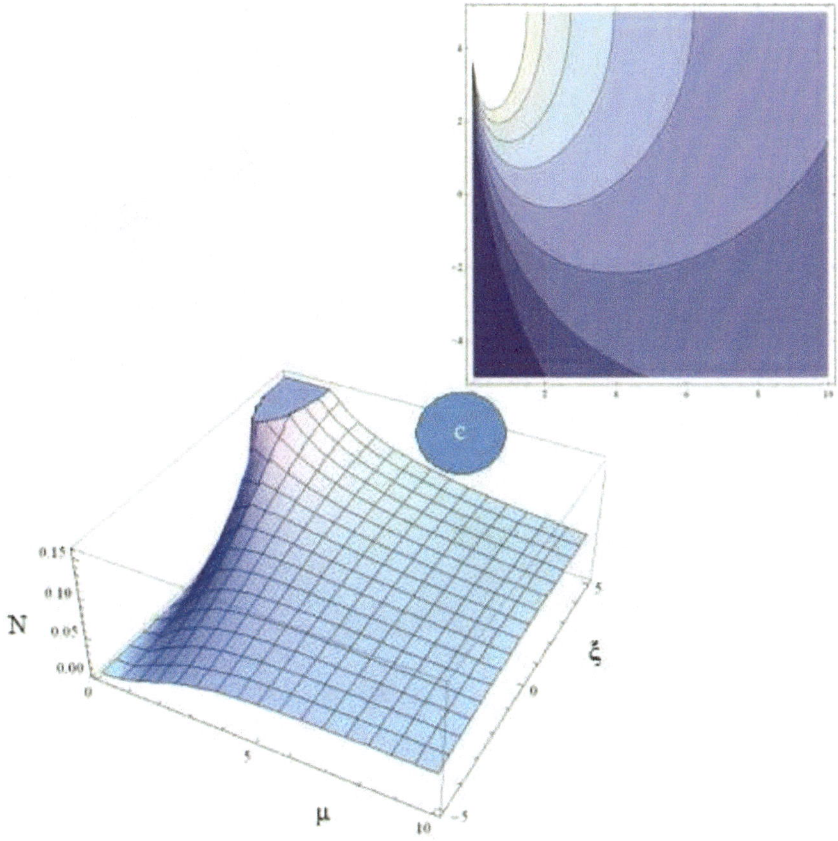

Fig. 6.16 Three- and two-dimensional (contour curves) the dependences $N(\xi, \tau = \text{const.}, \mu)$

$$< q_r^2 >^{1/2} \Delta r = m_0 \lambda (dt)^{(2/D_F)-1} \varepsilon_r(n, l), n = 1, 2, \ldots, l \le n - 1, \tag{6.76}$$

where:

$$\varepsilon_r(n, l) = \left\{ \left(1 - \frac{2l}{n} \frac{l+1}{n+1} \right) \left[\left(n^2 + 2 \right) - \left(\frac{l}{n} \right)^2 (l+1)^2 \right] \right\}^{1/2} \tag{6.77}$$

The function of states for this case is $\varepsilon_r(n, l) \ge \sqrt{3}$, in particular:

$$\begin{aligned} \varepsilon_r(n, l) &= \left(2 + n^2\right)^{1/2} \text{ for } l = l_{\min} = 0, \\ \varepsilon_r(n, l) &= \left(\frac{2n+1}{2n-1} \right)^{1/2} \text{ for } l = l_{\max} = n - 1. \end{aligned} \tag{6.78}$$

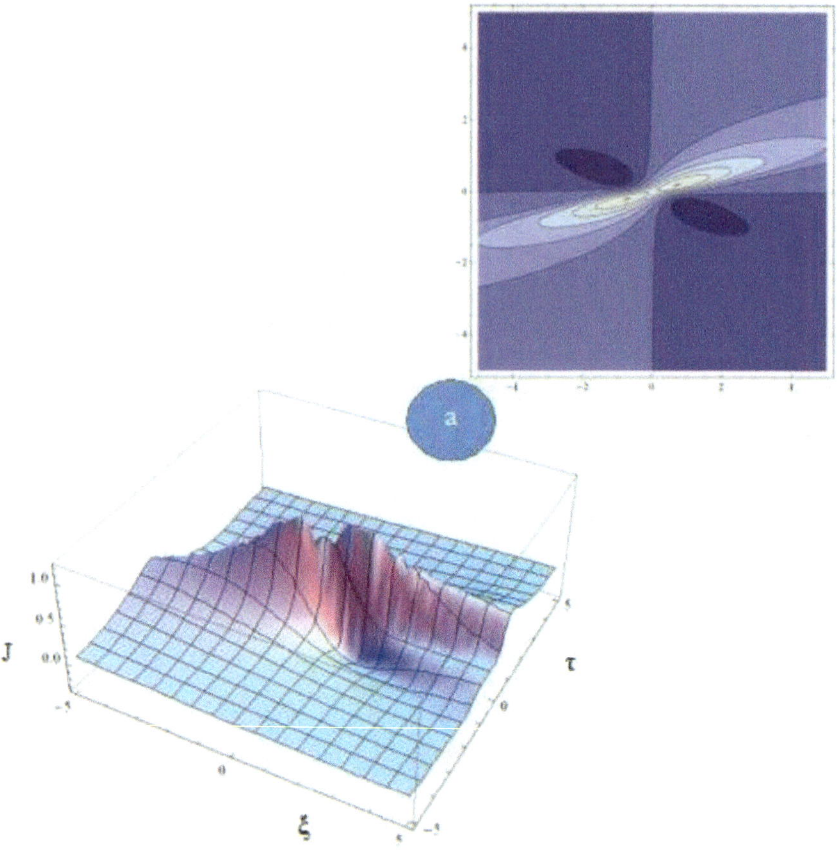

Fig. 6.17 Three- and two-dimensional (contour curves) the dependences $J(\xi, \tau, \mu = \text{const.})$

Similarly, $\varepsilon_\delta \neq 0$, $\varepsilon_{ij} = 0$ for $i \neq j$ and $\varepsilon_\phi = 0$.

(ii) By means of the solution for harmonic oscillator-type problems (one-dimensional), Eq. (6.21) verifies the following:

$$\begin{aligned} &< q_x^2 >= m_0^2 \lambda (dt)^{(2/D_F)-1}/\omega(2n+1), \\ &< (x- < x >)^2 >= \left[\lambda (dt)^{(2/D_F)-1}/\omega\right](2n+1). \end{aligned} \tag{6.79}$$

Accordingly, the uncertainty relation, Eq. (6.71), becomes

$$< q_x^2 >^{1/2} \Delta x = m_0 \lambda (dt)^{(2/D_F)-1}(2n+1), n = 0, 1, 2. \tag{6.80}$$

The function of states for this complex system is $\varepsilon(n) = 2n + 1 \geq 1$.

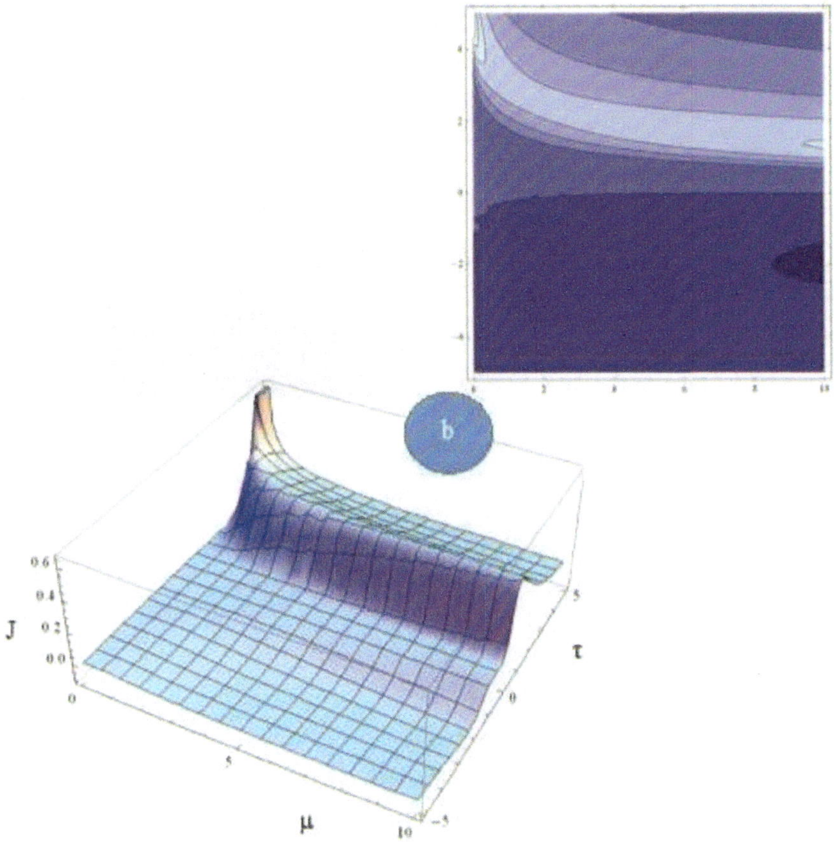

Fig. 6.18 Three- and two-dimensional (contour curves) the dependences $J(\xi = \text{const.}, \tau, \mu)$

(iii) Through the solution for the time-dependent (one-dimensional) motion of a free structural unit, Eqs. (6.41) and (6.42), one verifies that

$$< q_x^2 >= 2\lambda^2 (dt)^{(4/D_F)-2} \frac{1}{\mu(t)}$$
$$< (x- < x >)^2 >= \frac{1}{2}\mu(t). \tag{6.81}$$

Accordingly, the fractal uncertainty relation, Eq. (6.71), becomes

$$< q_x^2 >^{1/2} \Delta x = m_0 \lambda (dt)^{(2/D_F)-1}. \tag{6.82}$$

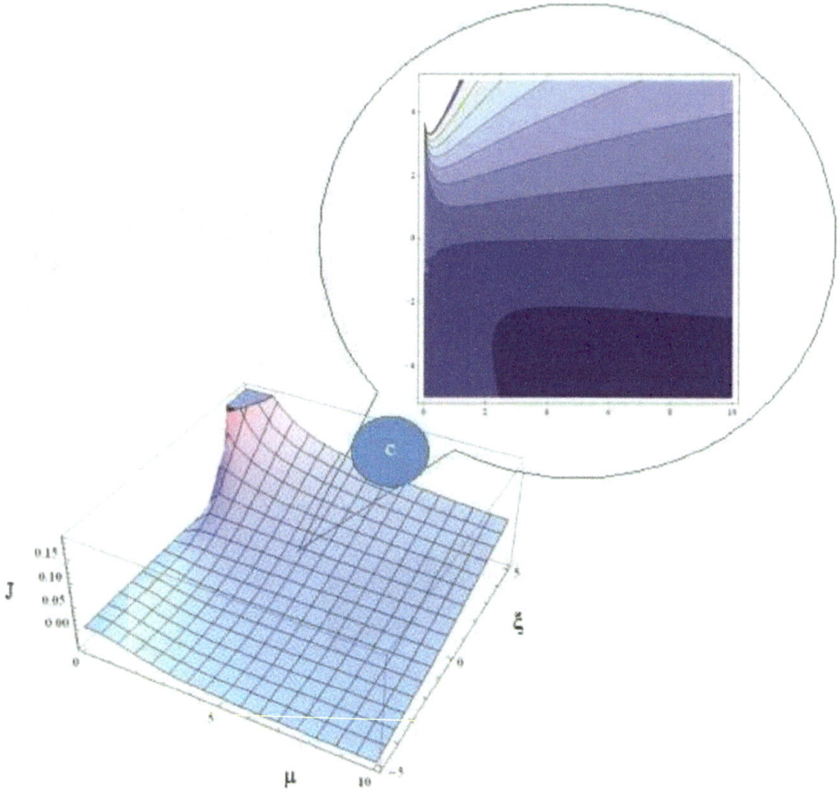

Fig. 6.19 Three- and two-dimensional (contour curves) the dependences $J(\xi, \tau = \text{const.}, \mu)$

Finally, a specific fractal momentum component (i) can be defined by the expression:

$$(2m_0 Q_i)^{1/2} \tag{6.83}$$

to form the uncertainty relation:

$$< 2m_0 Q_i >^{1/2} \Delta x_i = m_0 \lambda (dt)^{(2/D_F)-1} \varepsilon_i(s). \tag{6.84}$$

This uncertainty relation corresponds to that for the diagonal conjugate variables, Eq. (6.71).

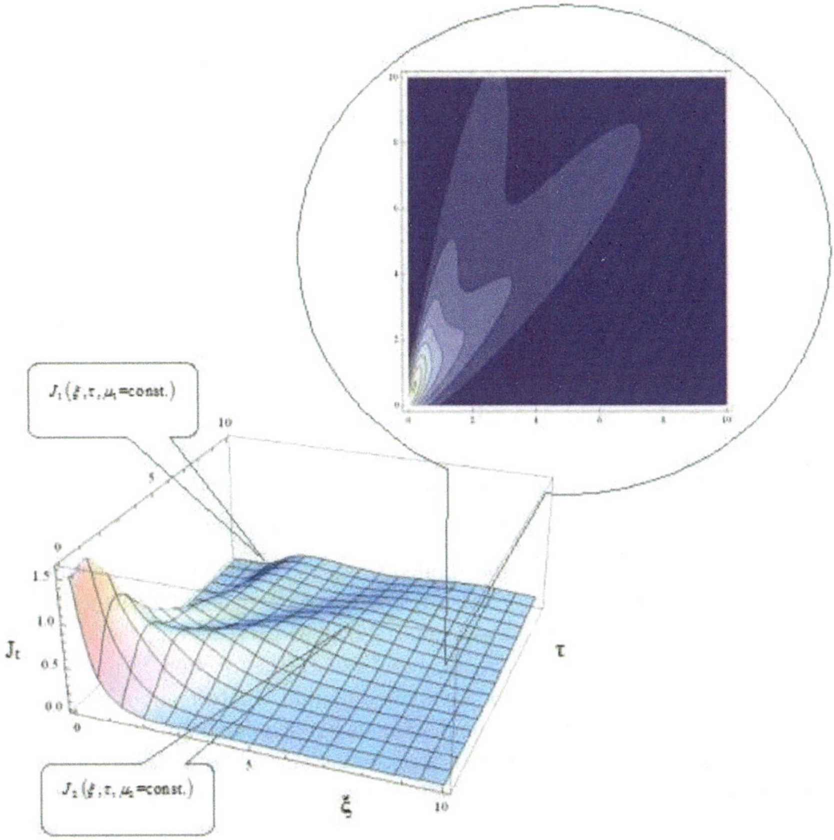

Fig. 6.20 Three- and two-dimensional (contour curves) dependences $J_1(\xi, \tau)$ at local scale $\mu_1 =$ const., $J_2(\xi, \tau)$ at local scale $\mu_2 =$ const. and $J_t(\xi, \tau)$ at the global scale $\mu_1 + \mu_2 =$ const. by means of the scale superposition principle

6.5 Wave-Particle Duality from the Perspective of Schrödinger- and Madelung-Type Scenarios

According to our previous results, the duality wave-particle model implies a fractal/multifractal medium that satisfies the constitutive law (5.38). Let us discuss this concept a little. In general terms, the constitutive law is a relationship between stresses and strains. Since our representations for these concepts are by matrices, a constitutive law is pure and simply a mathematical relationship—algebraic or analytical—between two 3X3 matrices. If $\boldsymbol{\sigma}$ is the stress matrix and $\boldsymbol{\varepsilon}$ is the deformation matrix, then a constitutive law is a relation of the form $\sigma = \sum \varepsilon$ where the matrix function is accessible for experimental evaluation or, in any case, evaluation of a different nature than that through purely algebraic calculation. Here, we insist on the

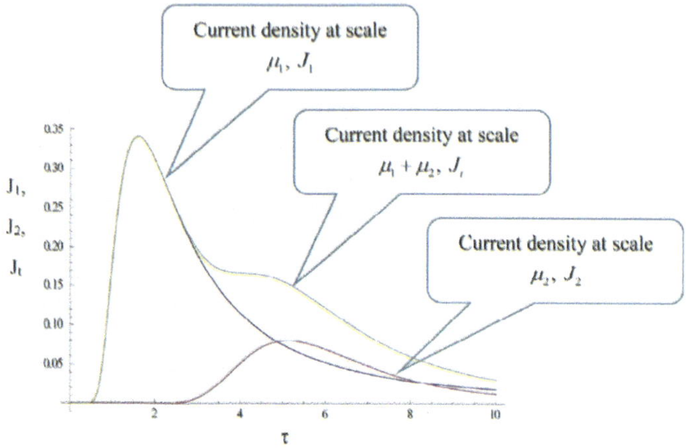

Fig. 6.21 Three- and two-dimensional (contour curves) dependences for $J_1(\tau)$ at local scale μ_1, $J_2(\tau)$ at local scale μ_2 and for $J_t(\tau)$ at the global scale μ_1, μ_2

meaning according to which $\boldsymbol{\sigma}$ is the applied stress and $\boldsymbol{\varepsilon}$ is the resultant strain. The reality referred to above then refers to the identity of the material characterized by the constitutive law. Indeed, in modern materials science, it is claimed that stress and strain matrices are universal mathematical tools, while the function $\sum$ is specific to the material to which the stresses $\boldsymbol{\sigma}$ are applied to induce deformations $\boldsymbol{\varepsilon}$. This can be seen in the concept of tension, which extends beyond applied force, a means of eliminating the force in general from the conceptual arsenal of mechanics. Only the tension applied from the outside is closely related to the idea of force. Otherwise, moving away from the idea of force, tensions can also be thought of as energy densities that characterize matter, occasionally even independent of any force. The central problem here is to find a function $\sum$ that implicitly contains the physical nature of the continuum to which the stresses and strains refer.

Now, upon a somewhat deeper reflection on the problem, a specific feature appears: here, we are dealing with uncontrollable manifestations. This is perhaps the main reason for maintaining the way of thinking from classical mechanics that considers force as a vector. It is very true that any action of ours is executed by force. In other words, we can only control forces and, if it can be anything else, only through forces. With rare exceptions, within the limits of mechanics, uncontrollable quantities are not explicitly taken into account because maintaining forces is a theoretically essential tool. Well, it just so happens that the most general idea of uncontrollability comes in very handy naturally with a constitutive law that we would call natural. Indeed, a constitutive law that relates the stresses to the strains must be of the form:

$$\boldsymbol{\sigma} = p_0\boldsymbol{e} + p_1\boldsymbol{\varepsilon} + p_2\boldsymbol{\varepsilon}^2, \tag{6.85}$$

where $\boldsymbol{e}$ is the 3×3 unit matrix. We call this equation the natural constitutive law, not because it would be realized in nature but because it naturally derives from our representations of stresses and strains. Indeed, our models for stresses and strains are 3×3 matrices, and if the constitutive law is analytic, Eq. (6.85) must automatically be true. Then, the relationship between two matrices can be represented in the form of a whole series in the deformation matrix, which is always reducible to one polynomial of the second degree according to the Hamilton-Cayley theorem. For the same reason, the relationship can be very well written with the interchanged locations of stresses and strains, so that deformations are still quadratic functions of stresses, only with other coefficients. Therefore, for a material, there are not only three characteristic numbers but also six: three for the expression of stresses in relation to the deformations and three for the expression of the deformations in relation to the stresses.

In this scheme, the material therefore has, at least apparently, a precise identity because we can identify it by the coefficients p_0, p_1, p_2, which are accessible to the experiment. This one is, in fact, what is usually meant by "characterizing the material". Too often, however, in experimental practice, these coefficients are considered properties of material purity, of the order of density or magnetism, but this restriction confuses concepts, sometimes with serious consequences, especially in engineering problems. Let us do this statement slightly more precisely. Regardless of what these material properties are, Eq. (6.85) shows that each of them can be extracted from experiments of loading a piece of material, either in extension or in compression. Furthermore, regardless of the nature of this loading, the main directions of the stresses coincide with the main directions of the deformations. On the other hand, if $\sigma_{1,2,3}$ are the main values of the stress matrix and $\varepsilon_{1,2,3}$ are those of the deformation matrix, then the constitutive law (6.85) is equivalent to the system

$$\begin{aligned}\sigma_1 &= p_0 + p_1\varepsilon_1 + p_2\varepsilon_1^2,\\ \sigma_2 &= p_0 + p_1\varepsilon_2 + p_2\varepsilon_2^2,\\ \sigma_3 &= p_0 + p_1\varepsilon_3 + p_2\varepsilon_3^2.\end{aligned} \tag{6.86}$$

For the sake of argument, we would be able to perform experiments that allow us to simultaneously measure all three main values of deformations and stresses. Such an experiment cannot be practically carried out, but the theoretical argument implies it always. The results of these experiments will allow us to calculate the properties of the material embodied in the coefficients $p_{0,1,2}$ from the system (6.86). Because the material is unique, we wait for a unique solution of the system, which is obtained only if its determinant

$$\begin{vmatrix} 1 & \varepsilon_1 & \varepsilon_1^2 \\ 1 & \varepsilon_2 & \varepsilon_2^2 \\ 1 & \varepsilon_3 & \varepsilon_3^2 \end{vmatrix} = (\varepsilon_2 - \varepsilon_3)(\varepsilon_3 - \varepsilon_1)(\varepsilon_1 - \varepsilon_2) \tag{6.87}$$

is nonzero. Thus, the parameters p_0, p_1, p_2 are indeed uniquely determined, regardless of the nature of the stresses imposed on the material if, and only if, the main deformations induced are different from each other. Regardless of the fact that they are unique and therefore very suitable for the characterization of materials, the coefficients thus obtained are not pure material properties because they depend on the stress state impressed on the material. We are therefore forced to specify still and further what we mean by pure material properties, and this is, as it has always been, a problem.

It is still a problem that can be solved by noting that the formalism shows that deformations occur even when there are no stresses applied to the material. Because we do not know their origin, these deformations can justifiably be taken as intrinsic properties, i.e., pure, of the material. However, let us call them only intrinsic, considering that they could be generated by forces whose presence we currently have no idea about. These can then be described by the following system of equations:

$$\begin{aligned} 0 &= p_0 + p_1\varepsilon_1 + p_2\varepsilon_1^2, \\ 0 &= p_0 + p_1\varepsilon_2 + p_2\varepsilon_2^2, \\ 0 &= p_0 + p_1\varepsilon_3 + p_2\varepsilon_3^2. \end{aligned} \tag{6.88}$$

Therefore, the intrinsic characterization of the material by experiment is now delegated to finding solutions for this linear and homogeneous system, if they exist. In fact, they always exist, all we have to decide is how many, and this fact depends on what we can measure in reality. If we always measure three different deformations in three orthogonal directions from space, then the material does not respond to the printed stresses. We already know this from our history: it is the primary quality of ether that propagates light instantaneously! However, because the simultaneous measurement of three main values for a matrix [14, 15] is only possible in our imagination, the quality of the material must be equally a product of our imagination. In reality, we can simultaneously measure at most two eigenvalues, which, when taken into consideration, shows us that the material could still respond to stresses; in other words, its deformation is truly accompanied by tension. Thus, if we measure one and the same value of the deformation in any direction in space, we have a double infinity of the stress states of the material, depending on two material parameters. If we measure two values of the deformation, in one direction and in its perpendicular plane, for example, then we have material stress states depending on a single material parameter. Assuming we can include one of the material parameters in a measured value, the most general constitutive law satisfied by the material presenting tensions accompanying the deformations will be of the form:

$$\boldsymbol{\sigma} = K(\boldsymbol{\varepsilon} - \varepsilon_1\boldsymbol{e})(\boldsymbol{\varepsilon} - \varepsilon_2\boldsymbol{e}), \tag{6.89}$$

where K is an arbitrary constant, which has the dimensions of a voltage. Such a material has three uncontrollable quantities, two of which are measurable.

In conclusion, we note that as long as we are interested only in measurable quantities, a convenient way to express the deformation matrix characteristics of a material that experiences uncontrollable deformation is in the form of a tensor:

$$\varepsilon_{ij} = \varepsilon_2\delta_{ij} + (\varepsilon_1 - \varepsilon_2) \cdot \mathrm{n}_i bfn_j i, j = 1, 2, 3, \tag{6.90}$$

where n is a unit eigenvector corresponding to the principal value ε_1. Such a material has distinctive directional properties in relation to the n direction, and these properties are given by the eigenvalues ε_1 and ε_2. In fact, Eq. (6.90) includes all the cases in the above view of the material if we agree to characterize its intrinsic properties as deformations. Note that this convention is independent of the constitutive description and must be guaranteed by our measurement capability. Therefore, we have this general conclusion: whenever the material deforms freely, i.e., under no perceptible force, its deformation matrix must be of the form (6.90) with all special cases included. The deformations, as well as the accompanying stresses, will then manifest as orthogonal tensors.

In the same way, we can discuss the category of materials capable of sustaining tensions without responding to deformations. This is, in fact, the essential property that matters to our senses. To express this, we must consider the opposite law, namely,

$$\boldsymbol{\varepsilon} = q_0\boldsymbol{e} + q_1\boldsymbol{\sigma} + q_2\boldsymbol{\sigma}^2. \tag{6.91}$$

This time, $\boldsymbol{\sigma}$ can only be abusively called tension; let us say, it represents the internal energy density of matter. Then, the defining state of this ether will be characterized by the system of equations

$$\begin{aligned} 0 &= q_0 + q_1\sigma_1 + q_2\sigma_1^2 \\ 0 &= q_0 + q_1\sigma_2 + q_2\sigma_2^2 \\ 0 &= q_0 + q_1\sigma_3 + q_2\sigma_3^2 \end{aligned}$$

which corresponds to the situation in which no deformation of the ether is observed. Again, the characterization of this ether depends on the number of nontrivial solutions of the system, and the most general form of the deformation matrix is here

$$\boldsymbol{\varepsilon} = K_1^{-1}(\boldsymbol{\sigma} - \sigma_1\boldsymbol{e})(\boldsymbol{\sigma} - \sigma_2\boldsymbol{e}), \tag{6.92}$$

where the constant K_1 has the dimensions of a voltage. Bell [16] reported that the relation (6.92) with $\sigma_1 + \sigma_2 = 0$ was a characteristic of metals, both in small and large deformations: metals always affect their hardness.

Again, as long as we are interested only in measurable quantities for the characterization of this material, its intrinsic stresses assume the following convenient tensor representation, analogous to Eq. (6.91):

$$\sigma_{ij} = \sigma_2\delta_{ij} + (\sigma_1 - \sigma_2) \cdot m_i m_j, i.j = 1, 2, 3, \tag{6.93}$$

where m is a unit vector corresponding to the main value σ_1. It can therefore be said with justification that the general property of the material that does not show deformations under tension is embedded in the form (6.92), all particular cases being included.

The case of Eqs. (6.90) and (6.93) is specific to the tensors that we would like to call here equivalent to a vector field. We understand this equivalence in the following way: let $\boldsymbol{v}$ be a vector field, with the help of which we build the matrix

$$v_{ij} = \alpha\delta_{ij} + \beta v_i v_j. \tag{6.94}$$

It is obvious that if v_k are the components of a vector and admitting α and β scalars, v_{ij} are automatically the components of a tensor. One of its main values, namely, α, is double. The other main value, which is different from α, is given by

$$\alpha' = \alpha + \beta v^2. \tag{6.95}$$

Note some interesting properties of this tensor. First, if either β or v_k is zero, v_{ij} is a purely spherical tensor. Second, if we calculate the eigenvector of $\boldsymbol{v}$ corresponding to the eigenvalue (6.95), we find that it is $\boldsymbol{v}$, up to a normalization factor. This property is independent of the parameter α, and in fact, it is what enables us to define the mentioned equivalence: given $\vec{v}$, we can directly construct $\boldsymbol{v}$ as a family of tensors depending on two arbitrary parameters that have this vector as an eigenvector. One may say that $\boldsymbol{v}$ represents some kind of action directed in the general direction of $\boldsymbol{v}$ but not exactly in that direction.

We note that the main advantage of the existence of such a constitutive law is that it allows us to express what we have just said about the fractal medium in space and the fractal medium in matter in an algebraic form. Then, it becomes apparent that the form of tensions that are not accompanied by deformations is characteristic of the case of a classic electric field, and the shape of the deformations that are not accompanied by voltages is characteristic of a classical magnetic field, or vice versa. However, as there is not only a fractal medium in matter or only a fractal medium in space, the true fractal medium must be a combination of those two species of fractal media. The simplest combination reveals a mathematical structure that describes electromagnetic radiation and therefore light.

A simple way to characterize the fractal medium is to admit that the fractal medium [17, 18] is characterized not by a single tensor of the shape (5.38) but by two, corresponding to two characteristic vectors, let us say $\mathbf{u}$ and $\mathbf{v}$. A tensor that would then describe the fractal medium could be of the form:

$$w_{ij} = \alpha\delta_{ij} + \beta u_i u_j + \gamma v_i v_j. \tag{6.96}$$

Somewhere along this line, the calculations are much more symmetrical if we write (6.96) in a more convenient form, namely,

$$w_{ij} = \lambda u_{ij} + \mu v_{ij}, \tag{6.97}$$

where λ and μ are real parameters that describe the "spatial" and "material" degrees, respectively, of the fractal medium, with the matrices ***u*** and ***v*** defined by

$$\begin{aligned} u_{ij} &= u_i u_j - \frac{1}{2}u^2\delta_{ij};\ v_{ij} = v_i v_j - \frac{1}{2}v^2\delta_{ij} \\ u^2 &= u_1^2 + u_2^2 + u_3^2;\ v^2 = v_1^2 + v_2^2 + v_3^2. \end{aligned} \tag{6.98}$$

Widely written, the matrix (6.97) is

$$w_{ij} = \lambda u_i u_j + \mu v_i v_j - \frac{1}{2}(\lambda u^2 + \mu v^2)\delta_{ij}. \tag{6.99}$$

It is then easy to see that this tensor has three real and distinct principal values. Indeed, its orthogonal invariants are

$$I_1 = -e;\ I_2 = -e^2 + g^2;\ I_3 = -e(e^2 - g^2), \tag{6.100}$$

where we put

$$e \equiv \frac{1}{2}(\lambda u^2 + \mu v^2);\ \vec{g} \equiv \sqrt{\lambda\mu}(\mathrm{u} \times \mathrm{v}). \tag{6.101}$$

The main values of w_{ij} can then be calculated as the roots of the characteristic equation of the matrix, and they are

$$w_1 = e,\ w_{2,3} = \pm\sqrt{e^2 - g^2}. \tag{6.102}$$

The pair in Eq. (6.101) is one of the eigenvectors of w together with its own value. The other two eigenvectors of w are perpendicular and located in the plane of the vectors **u** and **v**.

The quantities

$$w_n = \frac{w_1 + w_2 + w_3}{3},\ w_t^2 = \frac{1}{15}\left[(w_2 - w_3)^2 + (w_3 - w_1)^2 + (w_1 - w_2)^2\right] \tag{6.103}$$

are the averages for the normal and shear components of tensor ***w*** at any point. These quantities can be described, as we will see immediately, as components of a vector in a landmark characteristic of any point in space (regardless of whether he is in free space or in matter!) given by the eigenvectors of the matrix ***w***. We define the vector formed by the eigenvalues of the matrix ***w***:

$$|w\rangle \equiv \begin{pmatrix} w_1 \\ w_2 \\ w_3 \end{pmatrix}. \tag{6.104}$$

Now, if we choose the octahedral plane of the normal given by the unit vector

$$|n\rangle \equiv \frac{1}{\sqrt{3}} \begin{pmatrix} 1 \\ 1 \\ 1 \end{pmatrix} \tag{6.105}$$

the normal component w_n on this plane will be given by

$$\langle |w\rangle \equiv \frac{w_1 + w_2 + w_3}{3}. \tag{6.106}$$

The other quantity in Eq. (6.103) appears when we consider the norm of the tangent component from the octahedral plane, i.e., the norm of the vector:

$$|w_t\rangle \equiv |w\rangle - |n\rangle\langle n|w\rangle = \frac{1}{3} \begin{pmatrix} 2w_1 & -w_2 & -w_3 \\ -w_1 & +2w_2 & -w_3 \\ -w_1 & -w_2 & +2w_3 \end{pmatrix} \tag{6.107}$$

A simple calculation gives

$$\langle w_1 | w_1\rangle \equiv \frac{1}{5} w_1^2. \tag{6.108}$$

For the special case of eigenvalues (6.102), the two quantities become

$$w_n \equiv wn = -\frac{2}{\sqrt{3}} e, \; |w_1 = \frac{2}{3} \begin{pmatrix} -2e \\ 3\sqrt{e^2 - \vec{g}^2} + e \\ -3\sqrt{e^2 - \vec{g}^2} + e \end{pmatrix}. \tag{6.109}$$

As long as we claim that only the values (6.103) are measured, averaging the orientation of the vector in (6.109) in the octahedral plane remains undecided. This orientation is out of our control but can be measured against a reference direction in the plane octahedral. Let us assume that such a direction is given by a particular tensor ξ among those given in Eq. (6.98), with the characteristic vector $\boldsymbol{\xi}$. Repeating for this tensor as well the calculations above regarding the normal and shear component at that point in space, we have

$$\langle \xi \, |n\rangle = -\frac{1}{\sqrt{3}} \xi^2, \; |\xi_1 = \frac{2}{3} \xi^2 \begin{pmatrix} 2 \\ -1 \\ -1 \end{pmatrix} \tag{6.110}$$

When the vector $\boldsymbol{\xi}$ is perpendicular to both $\mathbf{u}$ and $\mathbf{v}$, the tensors $\boldsymbol{w}$ and commute. Therefore, we have a common reference system, and we can arrange ourselves so that the octahedral planes of these tensors coincide. In this case, the direction of the vector in Eq. (6.110) is fixed and can be taken as a reference in the octahedral plane. The angle ψ of the vector in Eq. (6.109) in relation to this direction is given by the usual geometric formula:

$$\cos\psi = -\frac{e}{\sqrt{4e^2 - 3\vec{g}^2}}. \tag{6.111}$$

Therefore, under the given conditions, ψ is independent of the reference vtor and depends only on the description of the fractal medium. By appropriately choosing the sign of the square root of the denominator of this expression, the origin $\psi = 0$ of that angle appears at e = g. In turn, this condition means that the angle θ between vectors $\mathbf{u}$ and $\mathbf{v}$ is given by the equation

$$|\sin\theta| = \frac{1}{2}\left|\frac{\lambda u^2 + \mu v^2}{\sqrt{\lambda\mu u v}}\right|. \tag{6.112}$$

As the quantity on the right side of this equation is always greater than or equal to unity, the angle between vectors $\mathbf{u}$ and $\mathbf{v}$ can only reach 90. Thus, the initial condition of the tensor $\boldsymbol{w}$. in the octahedral plane is translated by the fact that the vectors $\mathbf{u}$ and $\mathbf{v}$ are perpendicular between them, and their plane is perpendicular to the vector $\boldsymbol{\xi}$. If this last vector is given by the direction of a light ray, for example, we have here the classic image of light propagation after Fresnel.

References

1. Agop, M., Nica, P., Gurlui, S., Focsa, C., Magop, D., & Borsos, Z. (2011). The chaotic atom model via a fractal approximation of motion. *Physica Scripta, 84*, 045017.
2. Phillips, A. C. (2003). *Introduction to quantum mechanics*. Wiley.
3. Nottale, L. (1993). *Fractal space-time and microphysics: Towards a theory of scale relativity*. World Scientific.
4. Agop, M., Nica, P. E., Gurlui, S., Focşa, C., Păun, V. P., & Colotin, M. (2009). Implications of an extended fractal hydrodynamic model. Introduction of an extended fractal hydrodynamic model. *European Physical Journal D, 56*(3), 405419.
5. Mitchell, M. (2009). Complexity: A guided tour. Oxford University Press.
6. Cristescu, C. P. (2008). *Dinamici Neliniare şi Haos*. Fundamente Teoretice şi Aplicații (in Romanian), Romanian Academy Publishing House.
7. Dimitriu, D. G. (2013). Experimental and theoretical investigations of plasma multiple double layers. *Plasma Sources Science and Technology, 22*, 035007.
8. Agop, M., Dimitriu, D. G., Niculescu, O., et al. (2013). Experimental and theoretical evidence for chaotic dynamics of complex systems. *Physica Scripta, 87*, 045501.
9. Munceleanu, G. V., Păun, V. P., Casian-Botez, I., & Agop, M. (2011). The microscopic-macrosopic scale transformation through a chaos scenario in the fractal spacetime theory. *International Journal of Bifurcation and Chaos, 21*(2), 603–618.

10. Agop, M., & Casian-Botez, I. (2015). Theoretical approach of differential-non-differential scale transition in nanostructures. *JCTN, 12*, 1–12.
11. Agop, M., Forna, N., & Casian-Botez, I. (2008). New theoretical approach of the physical processes in nanostructures. *Journal of Computational and Theoretical Nanoscience, 5*(4), 483–489.
12. Colotin, M., Pompilian, G. O., Nica, P., Gurlui, S., Păun, V., & Agop, M. (2009). Fractal transport phenomena through the scale relativity model. *Acta Physica Polonica A, 116*(2), 157–164.
13. Timofte, A., Casian, B. I., Scurtu, D., & Agop, M. (2011). Systems dynamics control through the fractal potential. *Acta Physica Polonica A, 119*(3), 304311.
14. Mandelbrot, B. B. (1983). The fractal geometry of nature (Updated and augm. ed.), W.F. Freeman, New York.
15. Sorkin, R. (1994). Quantum mechanics as quantum measure theory. *Modern Physics Letters A, 9*, 3119–3127.
16. Bell, J. F. (1968). *An experimental study of instability phenomena in the initiation of plastic waves in long rods*. San Antonio University.
17. Lifshiëtìs, E. M., & Landau, L. M. (1987). *Fluid mechanics*. Pergamon Press.
18. Agop, M., Gavrilut, A., Crumpei, G., & Doroftei, B. (2014). Informational non-differentiable entropy and uncertainty relations in complex systems. *Entropy, 16*(11), 6042–6058.

Chapter 7
Chaos via Fractality in Gravitational Dynamical Systems

7.1 Introduction

As well known, the chaotic-type phenomena are relevant, in a large variety of science fields. Classical theory of physics defines the chaotic behavior as, an extreme sensitivity (hypersensitivity) to variation in the initial conditions of the equation describing the system dynamics. Subsequently, the systems that exhibit mathematical chaos are deterministic and thus ordered in some sense. Such nonlinear dynamical systems are, for example, the solar· system, the terrestrial atmosphere, turbulent fluids, the electronic generators and circuits [1–4]. However, in many systems where chaos arises, spatial and temporal structures seem also to appear, being frequently experiment, ally observed in chemistry (Belousov-Zhabotinsky reaction), electronics (Chua-Matsumoto circuit), fluid dynamics (Rayleigh-Bernard convention, Taylor-Couette flow), etc. These structures are in some cases determined or confirmed by means of numerical simulations, but very rarely understood or predicted from a fundamental theory [5]. The gravitational structures make no exception from the above-mentioned law. One of the most impressive examples is the fruitful suggestion made by Laskar that the inner planetary system (telluric planets) is chaotic [6]. It was also experimentally observed the regularity of the distribution of planets, satellites and asteroids in the solar system [7, 8].

In the following analysis, we propose a general method for approaching both chaos and pattern formation by presenting the nonlinear interaction between a neutral particle beam and a composite field (a gravitoelectromagnetic field overlapping a constant external gravitomagnetic field). Such a physical situation frequently appears in gravitational binary star systems [9, 10]. This chapter is structured in six sections. In paragraph 2, the gravitational Maxwell-type equations are presented. Further on, in paragraph 3, the equations of motion and their numerical solutions are obtained. The Lyapunov exponent analysis, bifurcation diagrams, and fractal analysis arc performed in Sects. 7.4, 7.5, and 7.6, respectively.

M. Agop and S. A. Irimiciuc, *Multifractal Theory of Motion*,
https://doi.org/10.1007/978-981-97-6410-5_7

7.2 Gravitational Maxwell-Type Equations

Consider the four-dimensional space–time as being split into space plus time (3 + 1). In this case, the general relativistic gravitational field (the contravariant form of the metric tensor) breaks into three parts [11, 13]:

1. An electric-like part, g^{00}, whose gradient for weak gravity is the Newtonian acceleration g_g;
2. A magnetic part, g^{Oi}, whose curl, for weak gravity, is the gravitomagnetic field $\boldsymbol{B_g}$;
3. A spatial metric, *gij,* whose curvature tensor is the "curvature of space".

The relativistic weak-field approximation starts with the postulate that the local coordinates, covering some patch on the space–time manifold, form an open neighborhood about the origin, in terms of which the metric is Minkowskian up to first order in coordinate over this patch. This construction may be carried out everywhere on the manifold.

In the weak (gravitational) field approximation, the metric tensor $g_{\mu\nu}$ can be written as

$$g_{\mu\nu} = \eta_{\mu\nu} + h_{\mu\nu} = \eta_{\mu\nu} + \overline{h}_{\mu\nu} - \frac{1}{2}\eta_{\mu\nu}\overline{h}, \tag{7.1}$$

where $\eta_{\mu\nu} = (+1,-1,-1,-1)$ is the usual Minkowski metric, $h_{\mu\nu}$ is a small perturbation of $\eta_{\mu\nu}$, and

$$h_{\mu\nu}\overline{h}_{\mu\nu} - \frac{1}{2}\eta_{\mu\nu}\overline{h}(\mu,\ \nu = 0,\ 1,\ 2,\ 3). \tag{7.2}$$

Within the same approximation, the contravariant form of the metric tensor writes

$$g^{\mu\nu} = \eta^{\mu\nu} - h^{\mu\nu};\ h^{\mu\nu} = \eta^{\mu\alpha}\eta^{\nu\beta}H_{\alpha\beta}; \tag{7.3}$$

while the determinant of $g_{\mu\nu}$ is

$$g = \det\left(g_{\mu\nu}\right) \approx -1 - h;\ (-g)^{\frac{1}{2}} \approx 1 + \frac{h}{2}. \tag{7.4}$$

Since all the covariant derivatives reduce to partial derivatives

$$\nabla_\mu h_{\alpha\beta} \approx \partial_\mu h_{\alpha\beta}. \tag{7.5}$$

The Riemann tensor $R^{\sigma}{}_{\mu\nu\lambda,}$ and the Ricci tensor $R_{\mu\nu}$ and the scalar curvature R write

$$R^{\sigma}_{\mu\lambda\mu} \approx \frac{1}{2}\left(\partial^\sigma\partial_\lambda h_{\mu\nu} + \partial\mu\partial\nu h^{\sigma}_{\lambda} - \partial_\mu\partial_\lambda h^{\sigma}_{\nu} - \partial_\nu\partial^\sigma h_{\lambda\mu}\right), \tag{7.6}$$

$$R_{\mu\nu} = R^{\lambda}_{\mu\lambda\nu} \approx \frac{1}{2}\left(\partial^{\lambda}\partial_{\lambda}h_{\mu\nu} + \partial_{\mu}\partial_{\nu}h_{h} - \partial_{\mu}\partial_{\lambda}h^{\lambda}_{\nu} - \partial_{\nu}\partial^{\lambda}h_{\lambda\mu}\right), \tag{7.7}$$

and

$$R = \eta^{\mu\nu}R_{\mu\nu}, \tag{7.8}$$

respectively.

Using Einstein's equations with the substitutions [12] and [13]

$$E^{t} = G^{00i} : B^{1} = G^{023}; B^{2} = G^{031}; B^{3} = G^{012}(i = 1, 2, 3), \tag{7.9}$$

where

$$\begin{cases} G^{\alpha\beta\gamma} = \frac{1}{4}\left(\overline{h}^{\alpha\beta,\gamma} - \overline{h}^{\alpha\gamma,\beta}\right) \\ E = -\nabla\Phi - \frac{\partial A}{\partial t}; B = \nabla \times A \\ \Phi = -\frac{1}{4}\overline{h}_{00}; A^{i} = \frac{1}{4}h^{0i} \end{cases} \tag{7.10}$$

and imposing the de Donder gauge conditions

$$\begin{cases} \overline{h}^{\alpha\beta}_{,\beta} = \left(h^{\alpha\beta} - \frac{1}{2}\eta^{\alpha\beta}h\right)_{,\beta} \\ h = \eta^{\alpha\beta}h_{\alpha\beta} = -\overline{h} \end{cases}. \tag{7.11}$$

The Maxwell-type equations for gravitoelectromagnetic field are obtained in the usual form:

$$\begin{cases} \nabla \cdot E = -\frac{\rho}{\varepsilon}; & \nabla \cdot B = 0. \\ \nabla \times E = -\frac{\partial B}{\partial t}; & \nabla \times B = -\mu j + \varepsilon\mu\frac{\partial E}{\partial t}. \end{cases} \tag{7.12}$$

In the previous relations E^{i} and B^{i} define the electric-type and magnetic-type components of the gravitational field (gravitoelectric and gravitomagnetic fields, respectively), c is the gravitational permittivity of vacuum

$$\varepsilon = \frac{1}{4\pi G}, \tag{7.13}$$

μ is the gravitational permeability of vacuum

$$\mu = \frac{4\pi G}{c^{2}}, \tag{7.14}$$

where j^{i} are the components of the mass current, G is Newton's constant, and c is the speed of light in vacuum [14]. This form of the decomposed equations is valid in any coordinate system. They are covariant under transformations of the spatial coordinates $x^i = x^i(x^1, x^2, x^3)$, but it should be stressed that, in general, once a particular coordinate system is chosen, the equations cannot be taken simply over another system of coordinates by an arbitrary coordinate transformation of the type $\bar{x}^{\sigma} = \bar{x}^{\sigma}(x^0, x^1, x^2, x^3)$. It is necessary to return to the original term of the decomposed equations and substitute the quantities according to the new system of coordinates.

Quantitatively, the linear approximation is very well justified on the surface of any member of our solar system. Even on the surface of a white dwarf, the deviation of g_{00} from unity does not exceed a value of the order of 10^{-5}. Therefore, we can postulate, ab initio, the Maxwell-type gravitational equation for linear gravitation.

7.3 Equations of Motion and Their Numerical Solutions

Let us consider the interaction between a beam of particles and a composite field, composed by a gravitoelectromagnetic field overlapping a constant external gravitomagnetic field. The mathematical model of such a dynamical system is described by the following equations:

(i) Maxwell's equations for the gravitoelectromagnetic field (7.12a–d).
(ii) Relativistic equation of motion for a single particle of the beam:

$$\frac{dr}{dt} = \frac{p}{m_0\gamma};\ \frac{dp}{dt} = m_0E + \frac{m_0}{\gamma}R \times (B + B_0). \tag{7.15}$$

In Eqs. (7.12) and (7.15) E and B are the gravitoelectric and, respectively, the gravitomagnetic field intensities generated by the beam, $\boldsymbol{B_0}$ is the constant and uniform gravitomagnetic field, $\boldsymbol{j} = n\, dr/dt$ and n are the current density and particle density of the beam, respectively, $\boldsymbol{R} = \rho/m_0$ is the rigidity (momentum per unit mass) and $\boldsymbol{\rho}$ is the relativistic particle momentum

$$\mathrm{p} = m_0\gamma \mathrm{v}, \tag{7.16}$$

with

$$\gamma = \left(1 - \frac{v_{\perp}^2}{c^2}\right)^{-1/2}. \tag{7.17}$$

Here m_0 is the rest mass of a single particle, $v_{\perp}$ is the component of the velocity orthogonal to $\boldsymbol{B_0}$, and γ is the relativistic Lorentz factor. In this application, we consider a field with the following components:

(i) A gravitoelectromagnetic field with harmonic space dependence of the form:

$$E(x, t) = \left(\mathrm{Re}\left(E_x(t)e^{ikx}\right), \mathrm{Re}\left(E_y(t)e^{ikx}\right), 0\right); \tag{7.18}$$

$$B(x,t) = \left(0, 0, \mathrm{Re}\left(B(t)e^{ikx}\right)\right), \tag{7.19}$$

where

$$\mathrm{k} = (k, 0, 0) \tag{7.20}$$

is the corresponding wave vector.

(ii) A stationary uniform gravitomagnetic field was chosen as

$$B_0 = (0, 0, B_0). \tag{7.21}$$

Using the dimensionless variables

$$\begin{aligned} T &= kct;\ P = \tfrac{p}{m_0 c};\ \gamma = \sqrt{1 + P_x^2 + P_y^2} \\ \mathcal{B}(T) &= \tfrac{B(t)}{kc} = \Omega_b; \\ \mathrm{E}_{x,y}(T) &= \tfrac{E_{x,y}(t)}{kc^2};\ \Omega_B = \tfrac{\omega_B}{kc} = \tfrac{\gamma\omega_c}{kc} = \gamma\Omega_c; \\ \Omega_P &= \tfrac{\omega_p}{kc} = \tfrac{1}{kc}\sqrt{\tfrac{nm_0}{\varepsilon}}, \end{aligned} \tag{7.22}$$

Equations (7.12c, d) and (7.15) then become

$$\frac{d\mathcal{B}}{dT} = -i\mathrm{E}_y, \tag{7.23}$$

$$\frac{d\mathrm{E}_y}{dT} = -iB + \frac{1}{\gamma}\Omega_P^2 P_y e^{-iX}, \tag{7.24}$$

$$\frac{d\mathcal{E}_x}{dT} = \frac{1}{\gamma}\Omega_P^2 P_x e^{-iX} \tag{7.25}$$

$$\frac{dX}{dT} = \frac{P_x}{\sqrt{1 + P_x^2 + P_y^2}}, \tag{7.26}$$

$$\frac{dP_x}{dT} = \frac{1}{\gamma}\left\{\Re e\left[\mathcal{B}(T)e^{iX}\right] + \Omega_B\right\}P_y + \Re e\left[\mathcal{E}_x(T)e^{iX}\right], \tag{7.27}$$

$$\frac{dP_y}{dT} = -\frac{1}{\gamma}\left\{\Re e\left[\mathcal{B}(T)e^{iX}\right] + \Omega_B\right\}P_x + \Re e\left[\mathcal{E}_y(T)e^{iX}\right], \tag{7.28}$$

where Ω_b is the dimensionless gyro-frequency of the gravitomagnetic field generated by the gravitoelectromagnetic field, Ω_B is the dimensionless cyclotron frequency generated by the stationary uniform gravitomagnetic field, Ω_c is the dimensionless relativistic cyclotron frequency generated by the stationary uniform gravitomagnetic field, and Ω_P is the dimensionless fluid "eigenfrequency". In order to obtain the values of parameters that define an efficient mechanism of acceleration, we suppose that the motion of a single particle of the beam occurs in the gravitoelectromagnetic field generated by the remaining particles of the beam. This way, the initial system with a self-consistent field is transformed into a system placed in external fields. We shall then consider the motion of a particle in a constant external gravitomagnetic field, in the presence of a transversal gravitoelectromagnetic field, i.e.,

$$B(T) = E_y(T) = He^{-iT};\ (H = const.) \tag{7.29}$$

$$E_y = 0;\ \Omega_P = 0. \tag{7.30}$$

Using a method developed in [15], the following equations describing the nonlinear dynamics of the model are obtained:

$$\frac{dX}{dT} = \frac{P_x}{\sqrt{1 + P_x^2 + P_y^2}}; \tag{7.31}$$

$$\frac{dP_x}{dT} = \Omega_c[\beta\cos(X - T) + 1]P_y;\ \beta = \frac{H}{\Omega_B} = \frac{|\Omega_b|}{\Omega_B}; \tag{7.32}$$

$$\frac{dP_y}{dT} = -\Omega_c[\beta\cos(X - T) + 1]P_x + H\cos(X - T). \tag{7.33}$$

Since analytical solutions to the system of Eqs. (7.31)–(7.33) are difficult to obtain, we shall integrate them numerically. Here are the elements of the complex regular motion, for $H = 0.06$, shown in Fig. 7.1. The numerical solutions of these equations and their correspondents with dynamics of the system are obtained by applying the fifth-order Runge–Kutta algorithm with an adaptive step size control. Let us initiate the numerical calculus for the parameter values $\Omega_B = 0.5$, $\gamma = 2$, and the initial values:

$$T = 0.001, X = 0.001, P_x = \sqrt{1.5}, P_y = \sqrt{1.5}. \tag{7.34}$$

In the following discussion we shall perform a complete and detailed nonlinear dynamical analysis of the system evolution described by (7.31–7.33): complete temporal series, Poincare sections, complete phase space, Lyapunov exponents, bifurcation diagrams, and fractal analysis Figure 7.1.

For small amplitudes of the gravitoelectromagnetic field (e.g., H = 0.06), the particle motion is complex but retains a regular character (Fig. 7.12a–g). Then, one can associate spiral-type "trajectories" to a periodical movement having large amplitude (Fig. 7.12a) and harmonic time-increasing momentum (Fig. 7.12b, c), either in two-dimensional phase space (Fig. 7.12d–f), or in three-dimensional phase space (Fig. 7.12g).

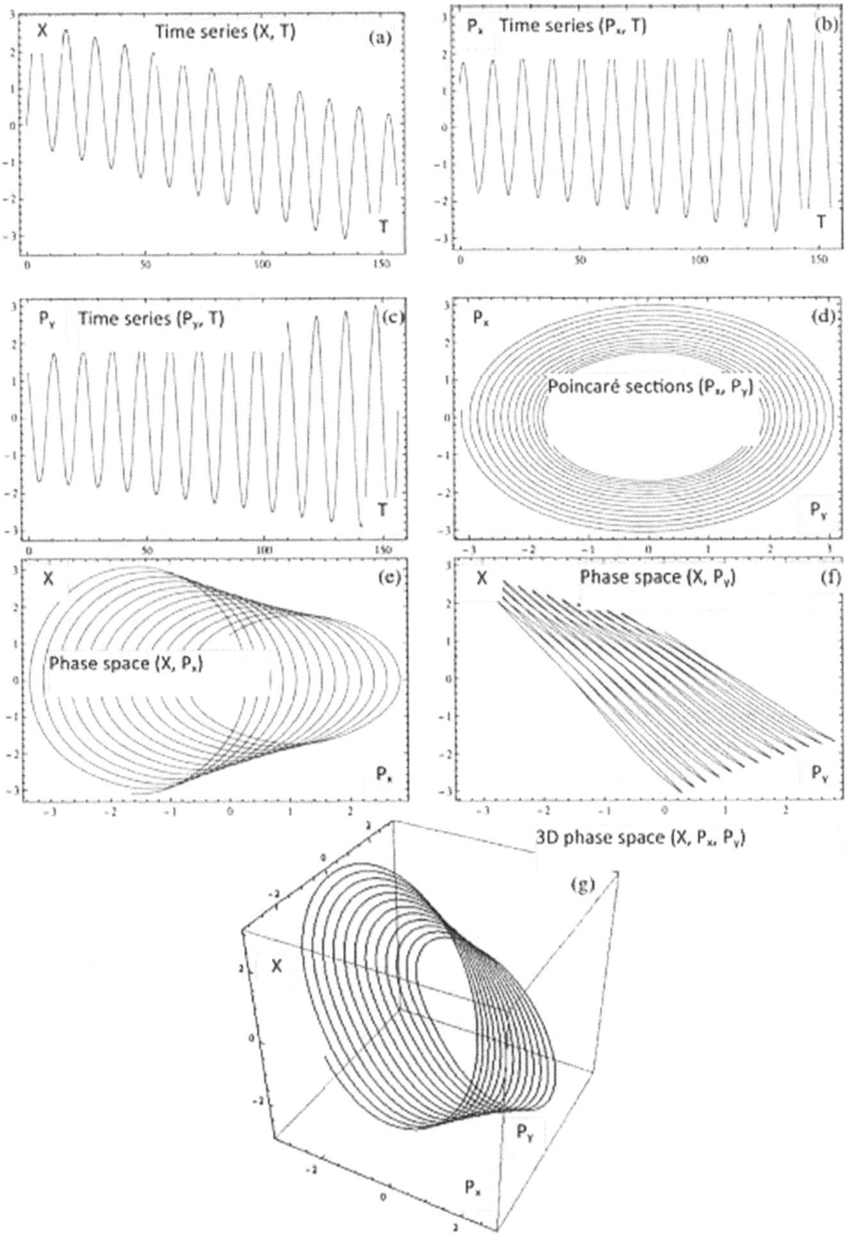

Fig. 7.1 The regular characteristic of the motion. **a** time series (X, T); **b** time series (P_x, T); **c** time series (P_y, T); **d** Poincare sections (P_x, P_y); **e** phase space (X, P_x); **f** phase space (X, P_y); **g** 3D phase space (X, P_x, P_y) for H = 0.06

The onset of fractalization (by means of stochastic processes) is observed once H exceeds 0.7 (see Fig. 7.2a–g). The debut of chaos through period doubling (Fig. 7.2b, c), as a natural result of the field-particle nonlinear interaction, initiates the particle rotation phase stochastic process in the considered gravitomagnetic field (for details see [15]). As a consequence, one can associate modified spiral-type "trajectories" to a periodic movement having small amplitude (Fig. 7.2a) and harmonic time increasing momentum, modified through period doubling (Figs. 26c), either in a two-dimensional phase space (Fig. 7.2d–f), or in a three-dimensional phase space (Fig. 7.2g).

Figure 7.2 shows the onset of fractalization by means of stochastic processes, for H = 0.7.

For H = 1.4, a gravitation gun-type effect is initiated (Fig. 7.3a–g). The stochastic expansion process induces the stochastic medium which is responsible for particle acceleration (the curves connecting the trajectories, either in a two-dimensional phase space or in a three-dimensional phase space, corresponding to the gravitation gun-drupe effect). The gun-effect is generated through the resonant field particle interaction. An extensive chaotic regime is obtained for H = 2.7 (Fig. 7.4a–g). Spontaneously and unpredictably, regular behaviors intercalate with chaotic ones (emphasized either through time series space (Fig. 7.4a, c) or through trajectories in a two-dimensional space (Fig. 7.4d, f) or a three-dimensional one (Fig. 7.4g). The system comes to an extended chaotic regime. A gravitational chaotic gun-type effect erupts for H = 3.49 (Fig. 7.5a–g) In the gravitational chaotic gun-type dynamics we distinguish the following sequences: at first, a localized chaotic regime emerges, subsequently a high-frequency oscillation with chaotic modulation of the amplitude appears and finally, we observe a sharp rectilinear part of the trajectory which in fact display the gravitational gun type effect (Fig. 7.5a, c) or through trajectories in a two-dimensional space (Fig. 7.5d, f) or trajectories in a three-dimensional space (Fig. 7.5g).

A gravitational multi-gun-type effect appears at H = 4.5 (see Fig. 7.6a, g). This is clearly illustrated by the time series (Fig. 7.6a–g) or through trajectories in a two-dimensional space phase or trajectories in a three-dimensional space. The gravitational multi-gun-type effect is correlated with the jumps between the different Larmor-type orbits. Vertical and large amplitude oscillations correspond to the Larmor-type orbits while the cures connecting them correspond to the gravitational chaotic gun-type effect.

Finally, we can comment on an interesting relation between the abstract formulas used and the associated physical reality. In this sense, a choice of type (29) for gravitoelectromagnetic field equation is compatible with certain cosmic structures for example with neutron stars. The gravitomagnetic field components of this structure is

$$B^G = 2\pi \frac{GM_N}{B_N T_N c^2} \cong 4.6 \times 10^2 s^{-1},$$

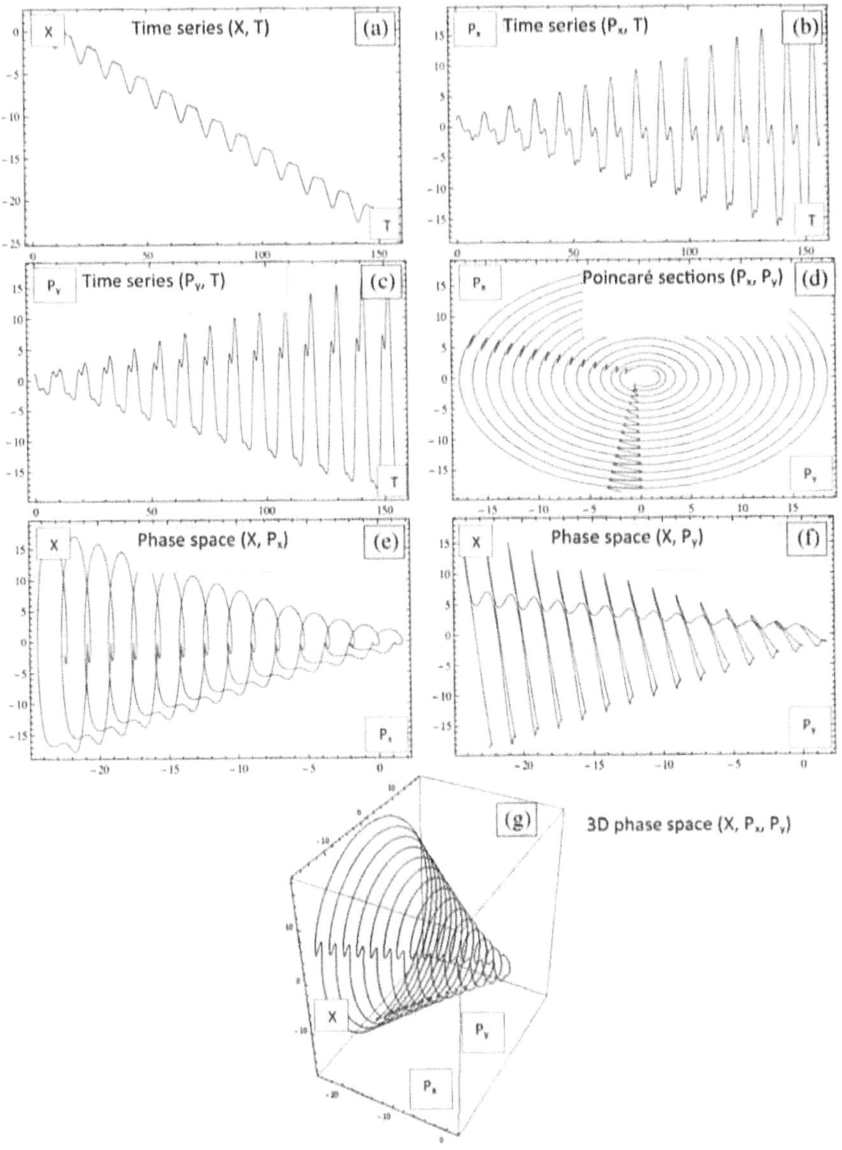

Fig. 7.2 The debut of chaos. **a** Time series (X, T); **b** time series (P_x, T); **c** time series (P_y, T); **d** Poincare sections (P_x, P_y); **e** phase space (X, P_x); **f** phase space (X, P_y); **g** 3D phase space (X, P_x, P_y) for H = 0.7

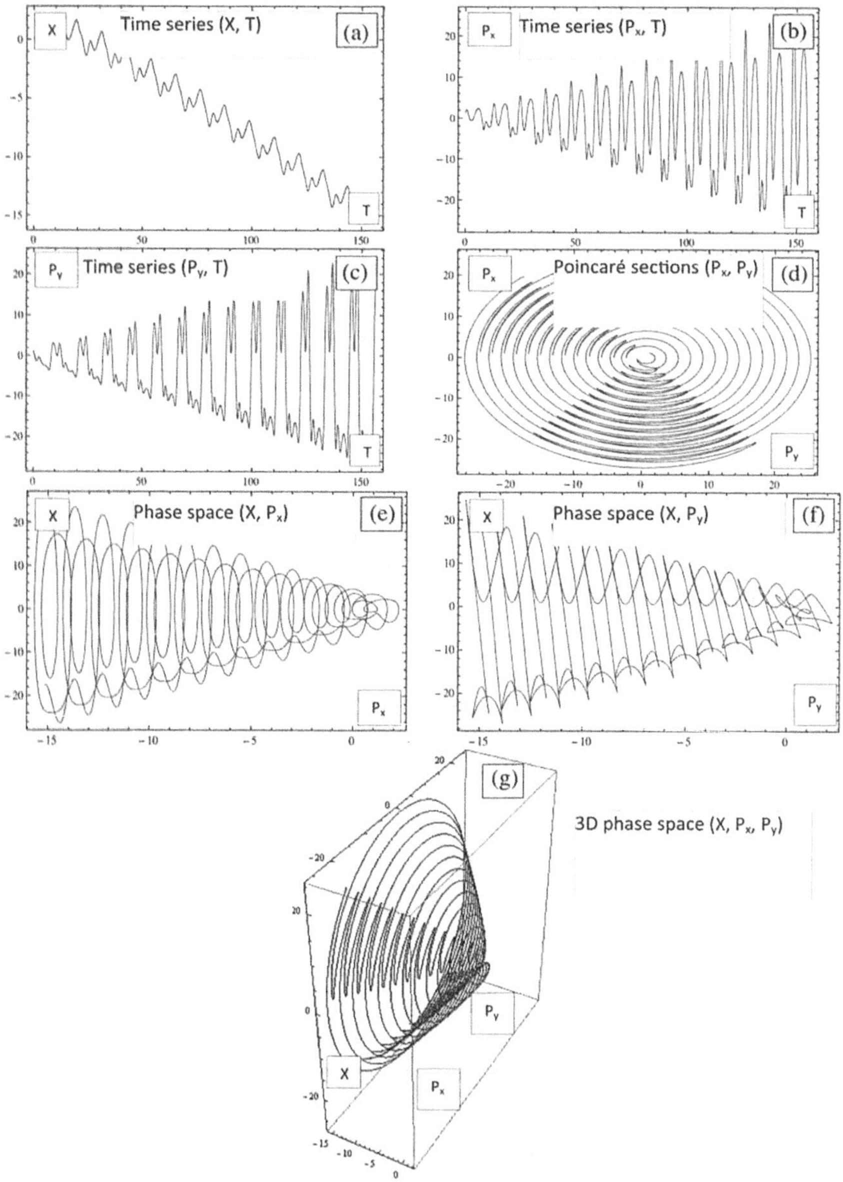

Fig. 7.3 The onset of a gravitational gun-type effect for H = 1.4: **a** time series (X, T); **b** time series (P_x, T); **c** time series (P_y, T); **d** Poincare sections (P_x, P_y); **e** phase space (X, P,); **f** phase space (X, P_y); **g** 3D phase space (X,P–c, P_y)

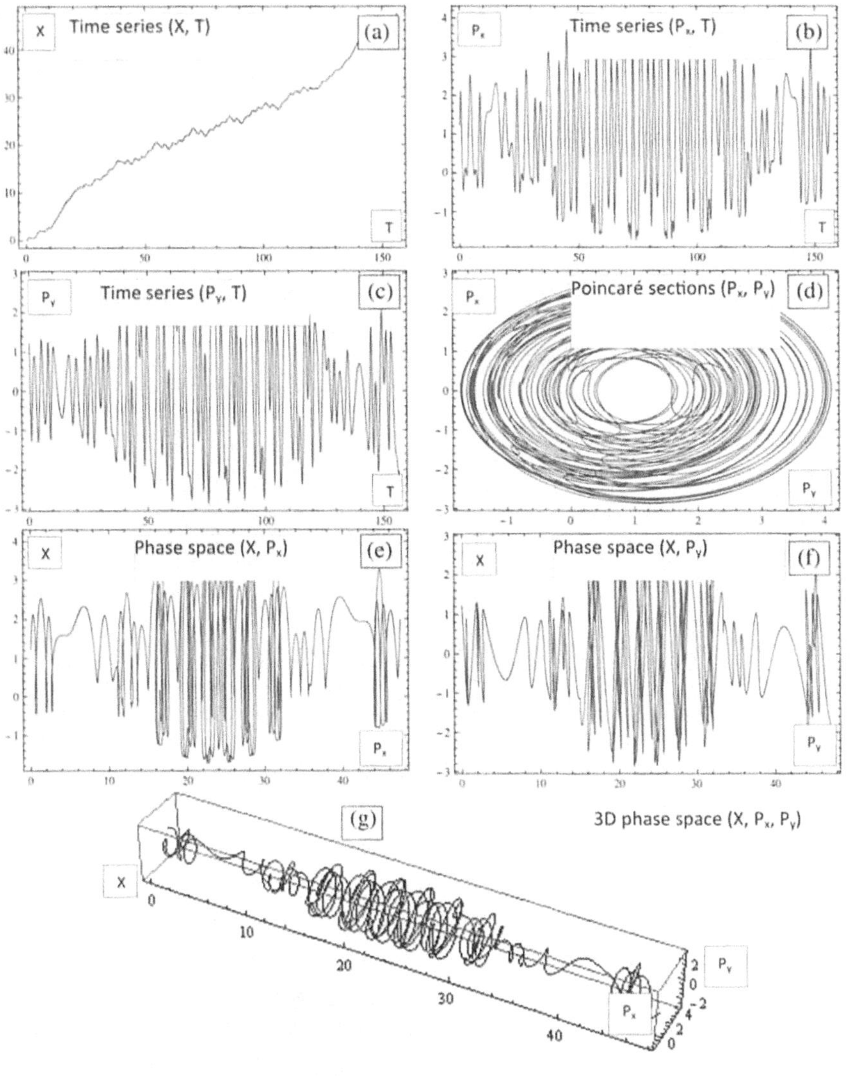

Fig. 7.4 Intermittent states for H = 2.7. **a** time series (X, T); **b** time series (P_x, T); **c** time series (P_y, T); **d** Poincare sections (P_x, P_y); **e** phase space (X, P_x); **f** phase space (X, P_y); **g** 3D phase space (X, P_x, P_y)

where $G = 6.6\ 10^{11}\ m^3kg^{-1}\ s^{-1}$ is Newton's constant and $c = 3\ 10^8\ ms^{-1}$ is the speed of light in vacuum. Let us choose the following values for the rest of the parameters: the star of the mass $M_N = 10^{30}$ kg, the star radius $R_N = 10^4$ m and the rotation period of the star 10^{-3} s. With these choices, the gravitoelectric field component becomes

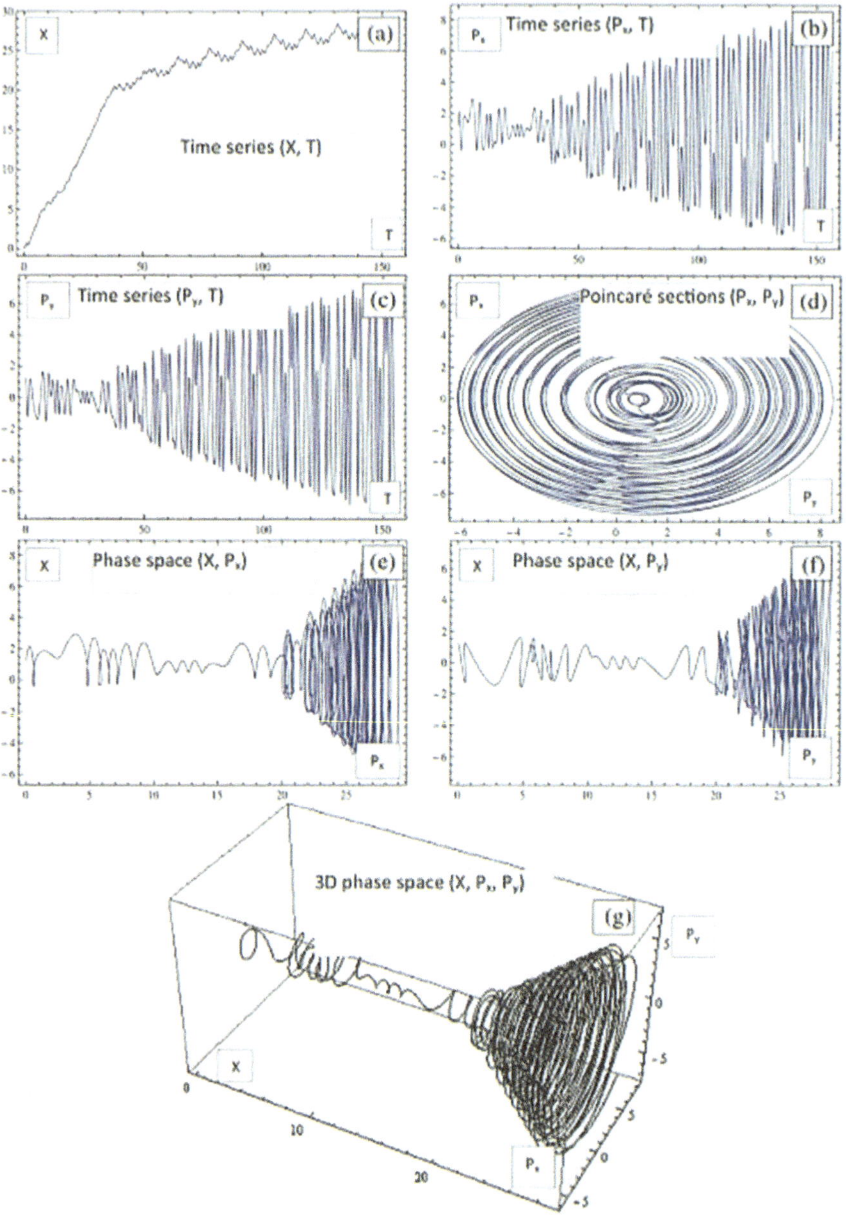

Fig. 7.5 Gravitational chaotic gun-type effect for H = 3.49. **a** time series (X, T); **b** time series (P_x, T); **c** time series (P_y, T); **d** Poincare sections (P_x, P_y); **e** phase space (X, P_x); **f** phase space (X, P_y); **g** 3D phase space (X, P_x, P_y) for H = 0.06

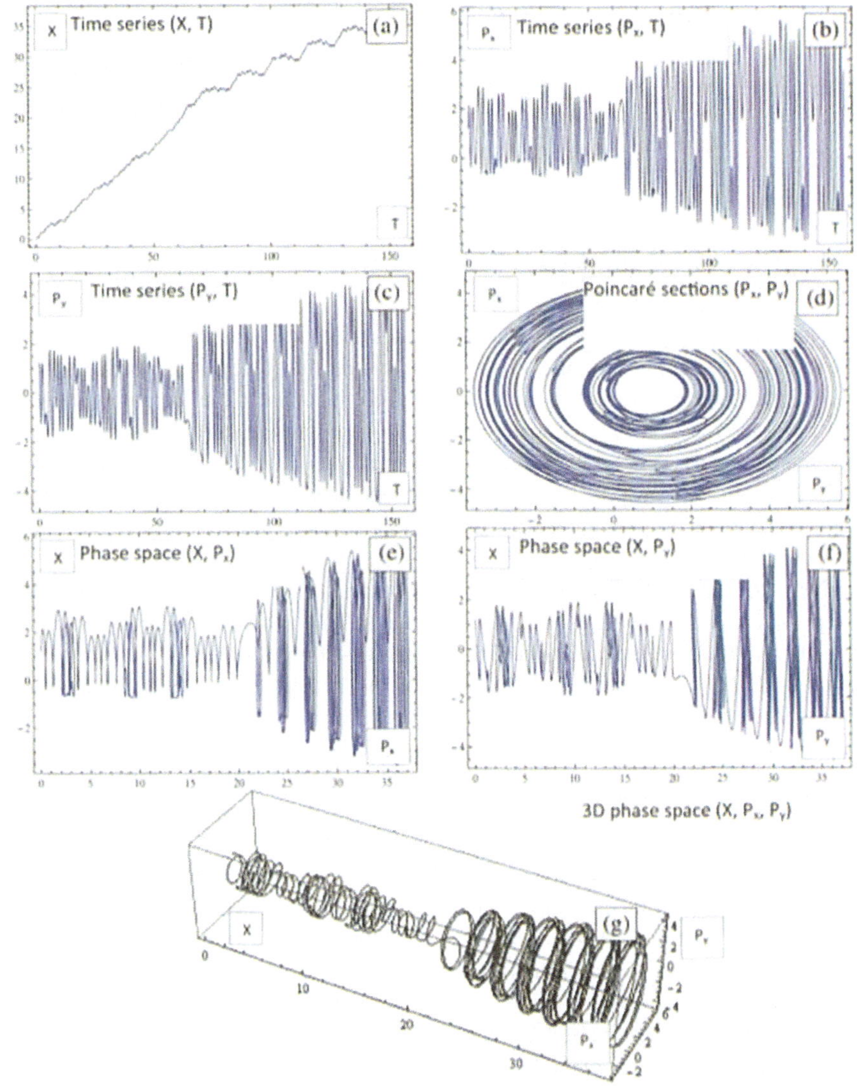

Fig. 7.6 Gravitational multi-gun-type effect for H = 4.5. **a** time series (X, T); **b** time series (P_x, T); **c** time series (P_y, T); **d** Poincare sections (P_x, P_y); **e** phase space (X, P_x); **f** phase space (X, P_y); **g** 3D phase space (X, P_x, P_y) for H = 0.06

$$E^G = \frac{GM_N}{R_N^2} \cong 6.6 \times 10^{11} ms^{-2}.$$

Under these circumstances, the relativistic beam particle is subjected to the force field:

$$F^G = cB^G \cong 1.4 \times 10^{11} ms^{-2},$$

which is comparable with E^G, at least for the order of magnitude. Therefore, this type of gravitoelectromagnetic field can be functional at the observable universal scale (e.g., white dwarfs).

7.4 Analysis of the Lyapunov Exponent

It is known the Lyapunov exponent dynamical system is a quantity that characterizes the rate of separation of infinitesimally close trajectories. Once the trajectories in the phase space and the laws of motion are analyzed, it comes to the idea of an increase of the value H = 2.5. By decreasing the observation interval starting from H = 4.75. We search for zones where the dynamics of our system enter a chaotic regime. It is difficult to directly observe an evolution towards chaos, because the form of circular trajectories does not change significantly (the particle returns only in some special cases).

Nevertheless, in the zone H = 4.75, one observes a successive de-doubling of the frequencies as precursors of the routes to chaos. This analysis has been performed by studying the Fourier transform of the signal for different values of the parameters. In order to have a more complete image of the chaotic zones, let us analyze the Lyapunov exponent in three cases, according to the modifications of initial conditions, for the three parameters X, P_x, P_y (see Figs. 7.7a–c). These graphs illustrate the two-dimensional mapping of the Lyapunov exponent as a function if the two main parameters H (along the x-axis) and Ω_B (along the y-axis). One can observe a strong similarity between these three graphs, as expected. Figure 7.7a–c, represents the contour diagrams for the Lyapunov exponent in terms of the parameters X, $P_{x,}$ and P_y, for which the initial conditions have been slightly modified. So, the values X(O) = 0.1, $P_x(0)$ = 2, and $P_y(0)$ = 1 have been considered at the initial moment. In each figure, the system evolution has been studied for small variations of about 0.1% for the initial conditions. Figure 7d illustrates an analysis of the chaotic zones. The darker areas represent the zones with higher Lyapunov exponent, i.e., the chaotic zones. We can thus confirm the fact that in order to obtain extensive chaotic regimes, we must have H > 2.5. Thus, three important regions are outlined, and islands 2 and 3 arc between 4 < H < 5, i.e., in the same zone resulting from the previous analysis.

In the above graphs, two variants of the color code are represented. In the first graph, the values for each Lyapunov coefficient level are shown, while in the second graph, the color code is being displayed (see Fig. 7.8). Between the brackets one can find the open intervals for λ. values, corresponding to each color. By means of a three-dimensional representation (see Fig. 7.9), one can observe a finer scale variation of the Lyapunov exponent. The stable zones (λ < 0) were eliminated from the graph, so that only positive values of the exponent were represented. Let us note that the Lyapunov exponent is a rate of chaos. More precisely, it is a quantitative measure of the sensitive dependence.

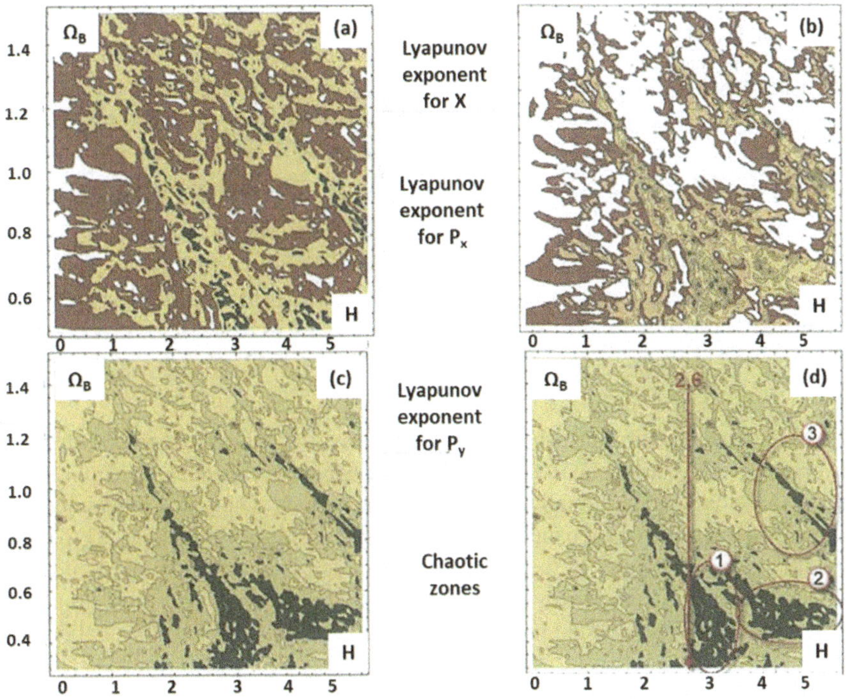

Fig. 7.7 **a–c**: The Lyapunov exponent in three cases, according to modification of the initial conditions for the parameters X, P, and P_y, respectively; **d** Analysis of the chaotic zones. The darker areas represent zones with higher Lyapunov exponent

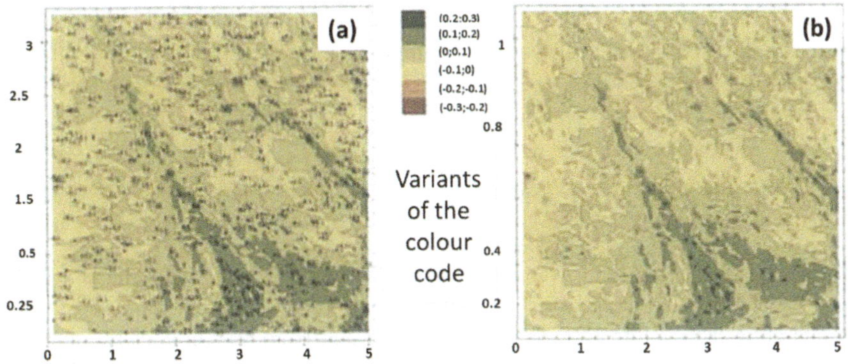

Fig. 7.8 **a, b**. Two variants of the color code (the scale of the Lyapunov coefficients)

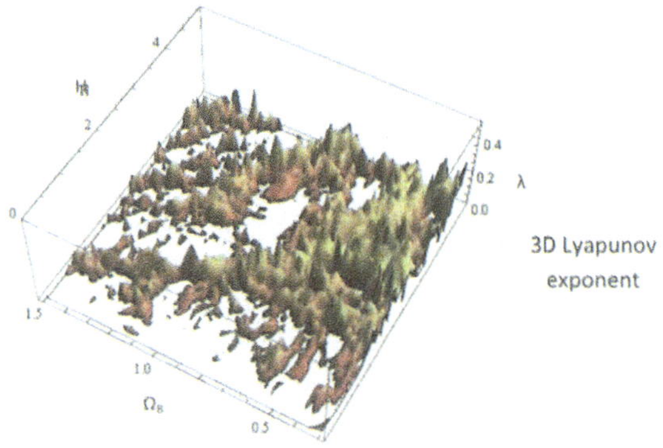

Fig. 7.9 A 3D representation of the Lyapunov exponent as a function of H and Ω_B

On the initial conditions, showing the rate at which, the trajectory converges or diverges in the phase space. The number of Lyapunov exponents is equal to the phase space dimension. Usually, the one with the highest value is the most important because, if it is positive, then the system is characterized by a high sensitivity to initial conditions in the vicinity of the initially considered point. Also, the original Chen-Lai algorithm gradually makes chaotic an arbitrarily given discrete-time dynamical system in terms of possessing positive Lyapunov exponents with uniformly bounded orbits [16]. There are several ways of defining the Lyapunov exponents [17].

Here we have used the following definition of the Lyapunov exponent where $\delta_{x0} = x_2(0) - x_1(0)$ is a small variation of the initial position, and $\delta_{x0}(t) = x_2(t) - x_1(t)$ is the distance between the positions of the system at the moment t. Within the frame of numerical analysis in relative units, we have considered $t/T = 10^3$, where T is the specific period of motion for $4 < \mathrm{H} < 5$, $\Omega_B = 0.5$ and $\delta_{x2}(0)/x_2(0) = 10^3$. Under these conditions, t is large enough and δ_{x0} is sufficiently small. Analysis for increasing precision has been performed, there wasn't a significant modification of the results.

7.5 Bifurcation Diagram

The scenario of evolution towards chaos through resonances overlapping is accomplished: stochastic layers (inside which the particle executes random walks) surrounding the cyclotron resonances are separated from one another by invariant curves. Some of these invariant curves may vanish (intersecting resonances) and a stochastic (fractal) web-like network arises in the phase portrait, where the layers do overall each other. The stochastic web may disappear quickly because of the strong nonlinearity of the physical system. The instability is illustrated by the fact that an elliptical point converts into a hyperbolic one. Two new elliptical points of the doubled period are then formed and this represents and island-doubling bifurcation. If the energy of the particle grows continuously a cascade of successive island-doubling bifurcations (a characteristic of the Hamiltonian Systems) occurs in the vicinity of elliptical points, which from necklaces of new, smaller, stability islands corresponding to higher order resonances in the interaction of the particle with the field. When the amplitude of the field increases, the stochastic energies can diffuse into the region of very high energies [18]. The bifurcation diagram confirms this scenario of transition to chaos (Fig. 7.10a–c).

We observe that, in order to obtain numerical values for bifurcation diagrams, we must find the frequencies for each evolution of X, P_x, or P_y as functions of H. To this end, we can use several methods. In our case, we did the mapping of X, P_x, or P_y versus time in a certain time interval. This time interval is swept along the time axis and the maxima are searched in this interval. If we find only one frequency of the same amplitude of oscillation, then, we found only one point. If we find two different frequencies, we find two different maxima and hence, the doubling of the frequencies appears. In essence, the method [18, 19] is based on the intermediate value theorem.

Only for inventory purposes we can mention here, among other efficient methods, the stochastic sensitivity function technique in the analysis of stochastic cycles and a

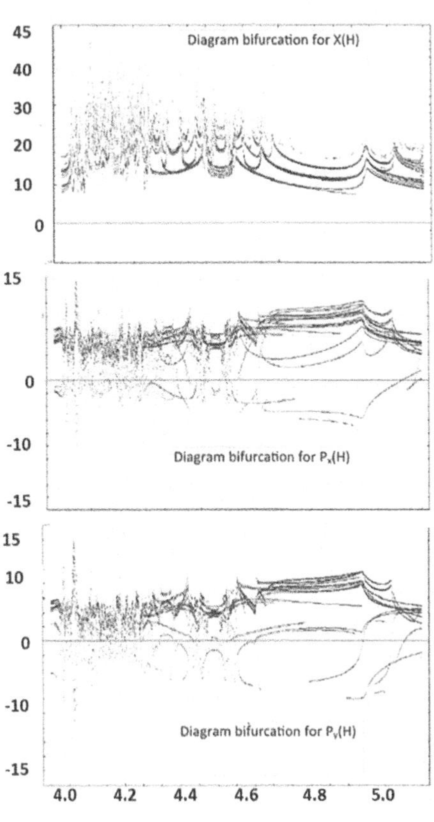

Fig. 7.10 The diagrams of bifurcation by means of resonances overlapping (for X(H)-(a), $P_x(H)$-(b), $P_y(H)$-(c)). The sequential chaotic domain corresponds to the vertical chaotic distributions of the points

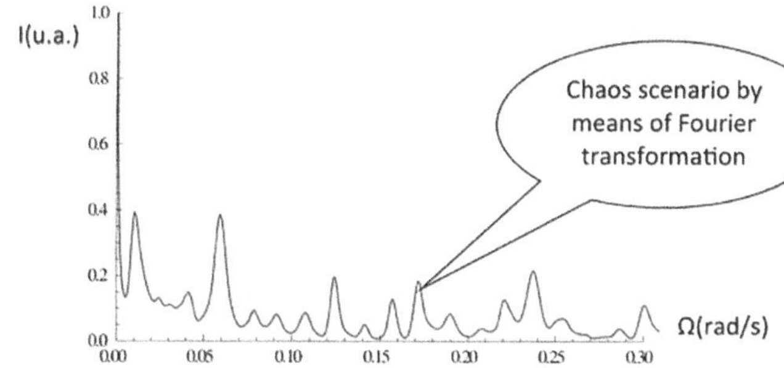

Fig. 7.11 Fourier transform specifies that the overlapping of the resonances takes place in the chaotic regime proximity

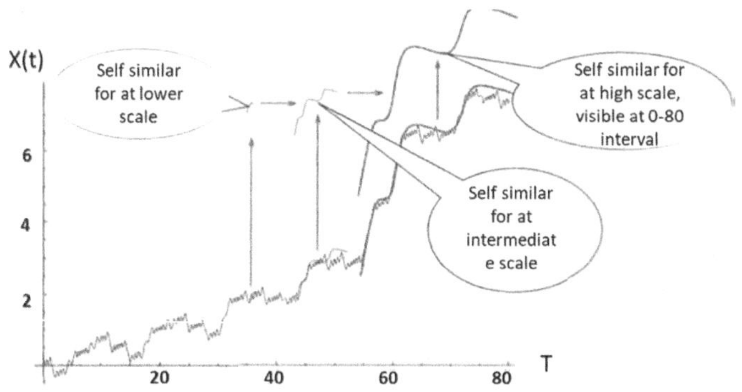

Fig. 7.12 The self-similarity of the trajectories at different resolution scales for $H = 4.471$, $\Omega_B = 1.52$, $\beta = 8.942$

new approach by constructing the dispersion ellipses of random trajectories for any Poincare sections [20]. Moreover, the Fourier transform specifies the same scenario, *i.e.*, the overlapping of these resonances takes place in the chaotic regime vicinity (see Fig. 7.11). We mention that, previous to chaotic behavior, the regime is linear and the Fourier transform can be applied. But, through the same transformation, the overlapping resonances scenario is inserted.

7.6 Multifractal Analysis

The dynamic analysis of paragraph 3 shows not only the chaotic behavior of the system, but also the fact that particle motion is found to be a fractal one. In a previous paper [21], we also show that the motion of particles takes place on continuous, but non-differentiable curves (fractal curves) with an arbitrary and constant fractal dimension. Our fundamental property of the fractal curve is self-similarity.

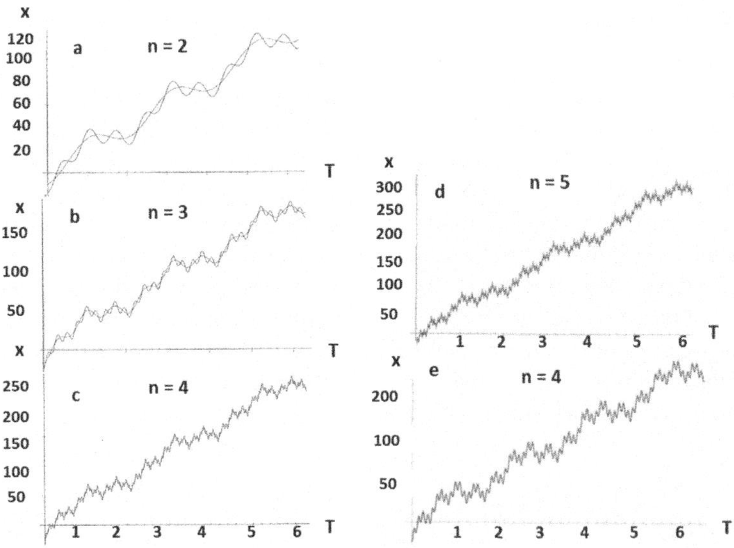

Fig. 7.13 Function (7.36) at different resolution scales in the presence of a guide (**a**)–(**d**), and in its absence (**e**)

The fractal analysis specific the self-similarity of the trajectory at different resolution scales, as shown in Fig. 7.12. It then follows that the trajectory can be approximated by a fractal. A function that approximates this trajectory is

$$X(T) = \sum_{i=1}^{n} \frac{-10\cos\left(3^{i}T\right)}{\sqrt{i}} + 10nT \tag{7.35}$$

In , this function is displaced at different resolution scales, either in the presence of a guide (a-d) or in the absence of it (e).

Considering that (7.36) is an alternative to the Weierstrass function, the fractal curve dimension is D = 1.5. A new study of nearly one million galaxies suggests matter in the universe is distributed in a fractal pattern (at any scale), with the fractal dimension going asymptotically to 1.5 [22]. We can conclude, therefore, that in the above investigations, a good coherence between the theoretical and experimental models is achieved. As a consequence, the trajectories of motion of the neutral particle as a result of its nonlinear interaction with a field are described by a fractal curve [23].

7.7 Conclusions

The nonlinear dynamics analysis (temporal series, Poincare sections, phase, space, Lyapunov exponents, bifurcation diagrams, etc.) of the particle-field nonlinear interactions (composite fields containing a gravitoelectromagnetic field and a static magnetic field) is extensively presented. Some physical mechanisms (gravitational gun type, chaotic gun type, and multi-gun-type effects) can well explain the transition of a test particle from one orbit to another (classical analog of quantum absorption). Fractal analysis shows that these curves possess the property of continuity and non-differentiability, i.e., they are fractal curves. In such a context we were able to apply the scale relativity theory to the study of the stationary states of the test particle, introducing the quantization at cosmological scale. Therefore, the trajectory fractalization represents a natural mechanism of quantization at a cosmological scale.

References

1. Strogatz, S. H. (2001). *Nonlinear dynamics and chaos: With applications to physics, biology, chemistry, and engineering*. Westview Press.
2. Mandelbrot, B. B. (2024). *Fractals and chaos: The Mandelbrot set and beyond*. New York: Springer.
3. Ott, E. (2002). *Chaos in dynamical system* (2nd ed.). Cambridge University Press.
4. Muthuswamy, B., & Chua, L. O. (2010). Simplest chaotic circuit. *International Journal of Bifurcation and Chaos, 20*, 1567–1580.
5. Nayfeh, A. H. (2000). *Nonlinear interactions: Analytical, computational and experimental methods*. New York: Wiley Interscience.
6. Laskar, J. (1989). A numerical experiment on the chaotic behavior of the solar system. *Nature, 338*, 237–238.
7. Sussman, G. J., & Wisdom, J. (1992). *Chaotic evolution of the solar-system Science, 257*, 55–62.
8. Laskar, J. (1994). Large scale chaos in the solar system. *Astronomy & Astrophysics, 287*, L9–L12.
9. Argyle, B. (2004). *Observing and measuring double stars*. Springer.
10. Iping, R. C., Sonneborn, G., Gull, T. R., Massa, D. L., & Hiller, D. J. (2005). Detection of a hot binary companion of eta Carinae. *The Astrophysical Journal, 633*, L37–L40.
11. Peng, H., & Qin, C. B. (1984). Precession of a gyroscope on the ground and the existence of magnetic-type gravitation. *Physics Letters A, 103*, 197–199.
12. Peng, H. (1990). A new approach to studying local gravitomagnetic effects on a superconductor. *General Relativity and Gravitation, 22*, 609–617.
13. Ciubotariu, C. (1991). Absorption of gravitational-waves. *Physics Letters A, 158*, 27–30.
14. Agop, M., Buzea, C. G., & Nica, P. (2000). Local gravity-electromagnetic effects on a superconductor. *Physica C: Superconductivity, 339*, 120–128.
15. Argyris, J., & Ciubotariu, C. (2000). A chaotic gun effect for relativistic charged particles. *Chaos Solitons & Fractals, 11*, 1001–1014.
16. Lai, D., & Chen, G. (2006). Distribution of controlled Lyapunov exponents via the Lai-Chen algorithm. *Computers and Mathematics with Applications, 52*, 1649–1656.
17. Elhadj, Z., & Sprott, J. C. (2009). Some explicit formulas of Lyapunov exponents for three-dimensional quadratic mappings. *Frontiers of Physics China, 4*, 549–555.
18. Antici, A., Marie, C., & Agop, M. (2009). Chaos through stochastization. Ars Longa, Iasi.

19. Mandelbrot, B. B. (1983). The fractal geometry of nature. New York: W.H. Freeman and Company
20. Gouyet, J. F. (1996). *Physics and fractal structures*. Springer.
21. Bashkirtseva, I., Chen, G., & Ryashko, L. (2010). Analysis of stochastic cycles in the Chen system. *International Journal of Bifurcation and Chaos, 20*, 1439–1450.
22. Munceleanu, G. V., Paun, V. P., Casian-Botez, I., & Agop, M. (2011). The microscopic-macroscopic scale transformation through a chaos scenario in the fractal space-time theory. *International Journal of Bifurcation and Chaos, 21*, 603–618.
23. Nottale, L. (2011). *Scale relativity and fractal space-time: A new approach to unifying relativity and quantum mechanics*. Imperial College Press.
24. Nottale, L. (2010). Scale relativity and fractal space-time: Theory and applications. *Foundations of Science, 15*, 101–152.

GPSR Compliance

The European Union's (EU) General Product Safety Regulation (GPSR) is a set of rules that requires consumer products to be safe and our obligations to ensure this.

If you have any concerns about our products, you can contact us on ProductSafety@springernature.com

In case Publisher is established outside the EU, the EU authorized representative is:

Springer Nature Customer Service Center GmbH
Europaplatz 3
69115 Heidelberg, Germany